POCKET

Factfile

of

BIRDS

Andromeda Oxford

Text and Maps: Jill Bailey

Project Editor: Graham Bateman
Designer: Frankie Wood
Cartographic Manager: Richard Watts
Proof Readers: Lin Thomas,
                 Marian Dreier
Production: Clive Sparling

Planned and published by
Andromeda Oxford Ltd
11–15 The Vineyard
Abingdon
Oxfordshire OX14 3PX

ISBN 1–871869–94–3

Origination by H.B.M. Print Ltd,
Singapore

Maps produced by Cosmographics,
England

Printed by Tien Wah Press, Singapore

CONTENTS

INTRODUCTION

Both birds and insects have achieved mastery of the air, but only birds are completely at home there: birds such as swifts (p. 76) and albatrosses (p. 11) spend almost their entire lives airborne, even sleeping and mating on the wing. Birds have several adaptations for their aerial lives. Their bones are hollow, which reduces their body weight. Feathers provide a warm, strong but very light covering: the weight of a bird is surprisingly small in relation to its apparent body size. Also, as the basis of the wings, feathers provide the means for flight. Their horny bill, scaly legs and feet, and the laying of eggs, are reminders of birds' reptilian ancestry.

Birds have been seen flying over the highest mountains on the planet, over the oceans a long way from land and across the most barren of deserts. They have managed to exploit most extreme environments, such as the Arctic tundra, by migrating vast distances to more favorable climates in winter. Birds have many different lifestyles. Tiny hummingbirds (pp. 199, 200) can hover almost stationary in front of flowers while they sip nectar with special brush-tipped tongues. Peregrine falcons (p. 68) can plunge-dive at speeds of up to 350 km (217 mi) per hour as they swoop on unsuspecting prey. Owls (pp. 70–75) hunt at night, using their acute hearing and eyesight; their extra-soft feathers muffle the sound of their approach. Ostriches, rheas and emus (pp. 90–93) have lost the ability to fly, but have very powerful legs and are fast runners. Their large size means that they have few enemies. Penguins (pp. 6–8) "fly" through water rather than the air. Many waterbirds and seabirds (pp. 6–53) are expert swimmers and divers, some even chasing their prey underwater.

About This Book

This pocket-sized book provides an overview of the bird world, with examples selected from most of the major bird groups and from all parts of the world. The **color artwork** illustrates each bird's appearance. For each bird illustrated, a **distribution map** displays the overall range of each species; for migratory birds the text defines the areas in which they are to be found at different times of the year. Land distribution is shown in green, that in open seas and oceans in dark blue and that in coastal areas red; where the distribution is very restricted a square has been placed over the area in question. The **text** gives important information about the kinds of habitat the bird lives in, its migration habits, size, form, diet, social life and breeding behavior, and conservation status, if it is threatened. Other fascinating facts about each species are also included in the text and captions. On the rare occasion where statistics are missing, this is because they are not known for certain.

Here birds have been divided into four groups based on lifestyles rather than relationships. *Waterbirds and Seabirds* include penguins, pelicans, albatrosses; ducks, geese and swans; terns and gulls; and other birds that spend at least part of their lives on water. *Hunters and Scavengers* include the vultures; the birds of prey and owls; insect-eaters, such as swifts, frogmouths, nightjars and bee-eaters; kingfishers and woodpeckers. *Large Ground Birds* are birds that seldom leave the ground. They include flightless birds, such as ostriches, rheas, cassowaries, emus and kiwis; fowls and game birds; and some more unfamiliar large birds. The final category of *Songbirds and Other Birds* is made up of examples of those bird groups not already described. Most are small- to medium-sized perching birds.

Abbreviations
L. length from bill to tail tip   WT weight

External Features of a Bird

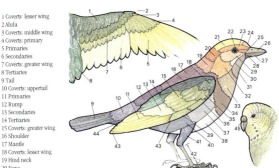

1 Coverts: lesser wing
2 Alula
3 Coverts: middle wing
4 Coverts: primary
5 Primaries
6 Secondaries
7 Coverts: greater wing
8 Tertiaries
9 Tail
10 Coverts: uppertail
11 Primaries
12 Rump
13 Secondaries
14 Tertiaries
15 Coverts: greater wing
16 Shoulder
17 Mantle
18 Coverts: lesser wing
19 Hind neck
20 Nape
21 Crown
22 Eye ring
23 Eyebrow
24 Forehead
25 Lores
26 Upper mandible
27 Lower mandible

28 Chin
29 Ear coverts
30 Mustachial stripe
31 Throat
32 Side of neck
33 Upper breast

34 Lower breast
35 Coverts: middle wing
36 Alula
37 Side
38 Flank (belly)
39 Coverts: primary wing

40 Flank
41 Thigh
42 Tarsus
43 Extent of remiges
44 Undertail coverts
45 Cere

EMPEROR PENGUIN

Aptenodytes forsteri

Family: Spheniscidae

DISTRIBUTION: Southern Ocean around Antarctica and the sub-Antarctic; breeds on sea ice or farther inland.

SIZE: L 112–115 cm (44–45 in); WT 22.7–45.3 kg (50–100 lb).

FORM: color black with white breast and undersides of wings; distinctive orange-yellow patch on side of neck, fading to yellow on throat; legs and feet black; bill black with orange streak; eyes dark.

DIET: mainly fish, prawns and cephalopods. Captures prey by pursuing it underwater, sometimes to depths of over 50 m (164 ft).

BREEDING: breeds in large colonies in winter. Does not build a nest. Male incubates single egg until it hatches

THE EMPEROR PENGUIN IS THE LARGEST PENGUIN. *It incubates its egg and warms its chick on its feet, where it is covered by a special fold of skin rich in blood vessels.*

after 60–66 days. Female now takes over while male returns to sea to feed. She feeds chick with regurgitated fish. After about 45 days, young penguins join a creche. They fledge at 5 months.

OTHER INFORMATION: usually walks upright on two legs, but will "toboggan" down slopes on belly.

CONSERVATION STATUS: not at risk, but some colonies are in decline. Breeding disturbed by human activity.

ADELIE PENGUIN

Pygoscelis adeliae

Family: Spheniscidae

A PAIR OF ADELIE PENGUINS *displaying after arriving at the breeding colony at the start of the breeding season. They may have walked up to 97 km (60 miles) from the sea to the nest site.*

DISTRIBUTION: Southern Ocean around Antarctica. Nests on ice-free rocky coasts or farther inland, at foot of ice cliffs.

SIZE: L 71 cm (28 in); WT 4.7–5 kg (10–11 lb).

FORM: color black with white belly and undersides of wings; legs and feet pinkish-gray; eyes dark, with white rings around them.

DIET: feeds close to shore, mainly on krill; also takes other crustaceans, fish and cephalopods. Captures prey by pur-suing it underwater, usually at depths of less than 20 m (66 ft).

BREEDING: in spring. Nest a small depression lined with pebbles. 2 eggs, incubated by both parents for 30–43 days. Young move to creche at 16–19 days and fledge after 50–56 days.

OTHER INFORMATION: when running or tobogganing, they can move faster than a human.

CONSERVATION STATUS: not at risk, but breeding is often disturbed near scientific bases, especially by over-flying.

JACKASS PENGUIN

Spheniscus demersus

African penguin,
Black-footed penguin
Family: Spheniscidae

UNLIKE MOST PENGUINS, THE JACKASS *penguin lives in quite a warm climate. It avoids the sun by fishing mainly at night, and by making its nest in burrows or under overhanging rocks or vegetation.*

DISTRIBUTION: waters off the coasts of South Africa and Namibia. Mostly breeds on inshore islands.

SIZE: L 68–70 cm (about 27 in); WT 2.96–3.31 kg (6.5–7.3 lb).

FORM: color black on back, head and wings, with white on belly extending up to side of head; narrow black horseshoe on breast; white patches on under-wings; legs and bill black, sometimes with pale band on bill.

DIET: feeds in inshore waters on small fish, crustaceans and squid, making dives as deep as 130 m (427 ft). Adults often fish communally.

BREEDING: all year in breeding colonies. Nests of plant material and small stones. 2 greenish eggs, rarely 4. Incubated by both parents for about 38 days. Young may join small creches. Fledge at 70–80 days. May breed twice each year.

OTHER INFORMATION: once decimated by egg-harvesting and collection of guano used to make fertilizer.

CONSERVATION STATUS: uncertain.

RED-THROATED DIVER

Gavia stellata

Red-throated loon
Family: Gaviidae

THE RED-THROATED DIVER HAS A *streamlined body for chasing fish underwater. Its feet are set well back to propel it along in the water, but this makes it walk rather awkwardly on land.*

DISTRIBUTION: breeds on tundra and moorland of the Arctic and near-Arctic, near a large lake or the sea. Spends winter in coastal waters farther south.

SIZE: L 53–69 cm (21–27 in); WT 1.15–1.98 kg (2.5–4.4 lb).

FORM: breeding plumage uniformly grayish-brown on upper parts, with white underparts, dark rust-red throat patch, vertical black-and-white stripes from nape of neck to crown; legs dark gray; bill paler gray; eyes reddish-brown. In winter finely spotted white.

DIET: mainly fish; also crustaceans and mollusks. Hunts underwater.

BREEDING: May to June. Nest ranges from simple scrape to heap of vegetation in shallow water. 2 olive eggs, with brown blotches. Eggs incubated by both parents for 24–29 days. Chicks leave nest and swim soon after hatching; usually stay under parent's wing or on its back. Fledge at 2 months.

OTHER INFORMATION: call a repeated throaty "kwuk".

GREAT CRESTED GREBE

Podiceps cristatus

Crested grebe,
Southern crested grebe (Australia)
Family: Podicipedidae

THE GREAT CRESTED GREBE HAS AN ELABORATE
courtship display, involving much head shak-
ing, preening and presenting of waterweeds,
and even running on water.

DISTRIBUTION: 3 main populations: a
circumpolar group, breeding beside
fresh or brackish lakes and ponds or
slow rivers in northern Europe, Russia
and North Africa, and wintering off
coasts farther south; scattered groups
across Africa; and further groups in
Australia and New Zealand.

SIZE: L 46–51 cm (18–20 in);
WT 0.6–1.49 kg (1.3–3.3 lb).

FORM: a slim, long-necked waterbird
with a flat head. Breeding plumage
dark brown; legs greenish; bill pinkish;
eyes crimson; double-horned crest on
head. Non-breeding adult paler.

DIET: mainly fish; also insects, cray-
fish, shrimps, mollusks, snails and
amphibians. Pursues prey underwater.
Hunts insects and spiders for young.

BREEDING: usually in spring in temperate climates, nonseasonal in tropics.
Nest a platform of rotting vegetation at
edge of water, sometimes floating. 3–6
white eggs, incubated by both parents
for 25–31 days. Fledge at 71–79 days.
Sometimes 2 broods a year.

OTHER INFORMATION: lives alone or in
loose groups. Call a harsh honk or bark.

ROYAL ALBATROSS

Diomedea epomophora

Family: Diomedeidae

DISTRIBUTION: breeds around coasts of southern New Zealand, and on Campbell Island, Auckland Island and Chatham Island. Disperses widely over Southern Ocean after breeding, as far as South America.

SIZE: L 107–122 cm (42–48 in); WT 8.2–9.0 kg (18–20 lb).

FORM: a large white seabird. Wings have black trailing edges; bill long, yellowish with black cutting edges on upper mandible.

DIET: mainly squid; some crustaceans and fish. Often feeds at night.

BREEDING: every other year, in spring.

THE ROYAL ALBATROSS SPENDS MOST OF ITS LIFE AT sea. Its long, narrow wings help it to soar over the ocean waves, snatching fish and squid from the surface waters.

Nest a pile of grass on the ground, surrounded by a rim of mud or a ditch. Several pairs nest close together. 1 whitish egg, incubated by both parents for 79 days. Chicks brooded for 4–5 weeks. Fledge after about 240 days. Reach sexual maturity at 9–11 years.

OTHER INFORMATION: can live up to 80 years. Wingspan up to 351 cm (11.5 ft).

CONSERVATION STATUS: not at risk, but declining slowly. Main threats are human disturbance while breeding, and tangling in fishing tackle.

GREATER SHEARWATER

Puffinus gravis

Family: Procellariidae

DISTRIBUTION: breeds in the South Atlantic, on Tristan da Cunha, St Helena and the Falkland Islands, on coastal areas of tussock grass or woodland. Then migrates around the North Atlantic to northeast Canada, then Great Britain and Iberia and on back to its southern breeding grounds.

SIZE: L 43–51 cm (17–20 in); WT 715–950 g (1.6–2.1 lb).

FORM: a narrow-winged gull-sized seabird that skims over the waves. Color dark brown with darker cap and white underparts that extend to throat and cheeks; vague dark patch on belly; white patch across base of tail; legs gray and pink; bill black; eyes brown.

THE GREATER SHEARWATER IS A BIRD OF THE OPEN *ocean. It can often be seen following ships to feed on fish offal. Large numbers gather on the sea near a breeding colony, waiting until nightfall to visit their chicks.*

DIET: mostly fish and squid; some crustaceans. Mostly plunge-dives on prey from height of up to 10 m (33 ft); also snatches fish from surface, or chases them underwater.

BREEDING: in spring. Forms large breeding colonies. Nests in burrows or among boulders in long grass. Single white egg incubated by both parents for 53–57 days. Chicks fledge at 105 days.

OTHER INFORMATION: wingspan up to 100 cm (39in).

SNOW PETREL

Pagodroma nivea

Snowy petrel
Family: Procellariidae

DISTRIBUTION: the Southern Ocean around the pack ice of Antarctica and the sub-Antarctic islands. Breeds on cliffs at altitudes up to 2,400 m (7,900 ft), sometimes as far as 325 km (200 mi) inland. Seldom travels far from nesting colony.

THE GHOSTLY APPEARANCE OF THE SNOW PETREL *fluttering over the water, dipping into it from time to time to seize prey, tells sailors that they are close to the pack ice.*

SIZE: L 30–40 cm (12–15 in); WT 240–460 g (8.5–16.2 oz).

FORM: a pure white bird with a very small black bill; legs dark bluish-gray; eyes brown.

DIET: mainly crustaceans, especially krill; also fish, squid and carrion such as whale and seal carcasses.

BREEDING: in spring. Forms breeding colonies, nesting in rock crevices. The single white egg is incubated by both parents for 41–49 days. Chicks brooded for about 8 days, and fledge at 41–54 days at which stage the chicks are often heavier than the parents.

OTHER INFORMATION: may live up to 20 years. Seldom alights on the water; rests on ice floes. On nests, discourage intruders by spitting or regurgitating foul-smelling oil.

COMMON DIVING-PETREL

Pelecanoides urinatrix

Falkland diving-petrel, Berard's diving-petrel, Kerguelen diving-petrel, Subantarctic diving-petrel, Tristan diving-petrel
Family: Pelecanoididae

THE COMMON DIVING-PETREL IS A SMALL *seabird that pursues its prey under water after plunging from a height or diving from the surface.*

DISTRIBUTION: sub-Antarctic and temperate regions of the Southern Ocean, from southern Australia and New Zealand to south Georgia, the Falkland Islands and Tristan da Cunha. Breeds among tussock grass on steep slopes of oceanic islands, sometimes inland.

SIZE: L 20–25 cm (8–10 in);
WT 105–165 g (3.7–5.8 oz).

FORM: color black, with gray cheeks and white underparts; underwings gray-tinged; legs blue; eyes dark brown.

DIET: mainly small crustaceans.

BREEDING: season varies according to location. Breeds in colonies with very

high density of burrows. Nests in burrows up to 150 cm (5 ft) long, without nesting material. Single white egg is incubated by both parents for 53–55 days. Chicks fledge at 45–59 days. Sexual maturity at 2–3 years.

OTHER INFORMATION: silent at sea, but makes a range of sounds, from mewing and cooing to wailing and chattering over breeding grounds. Has a rapid, fluttering flight. Uses wings as flippers.

GREAT WHITE PELICAN

Pelecanus onocrotalus

Eastern white pelican, Old World white pelican,
European pelican, Rosy pelican
Family: Pelecanidae

DISTRIBUTION: breeds in reedbeds and marshes from eastern Europe to western Mongolia, northwest India, Nepal and Bangladesh, and Africa south of the Sahara. Wintering areas include northeast Africa, Middle East and Pakistan.

THE GREAT WHITE PELICAN OFTEN FISHES IN A GROUP *in shallow water. The pelicans form a horseshoe shape to drive fish into a ball in the middle. Then they simultaneously dip their heads into the water and scoop up the fish.*

SIZE: L 148–175 cm (58–68 in); WT 5.4–15 kg (12–33 lb); male larger than female.

FORM: color white with black wingtips; yellow patch on breast; legs pink or orange; bill blue and pink; throat pouch yellow or pink; eyes dark. In breeding season develops crest and pinkish tinge.

DIET: mainly fish; also young birds.

BREEDING: spring in temperate zones, nonseasonal in tropics. Forms large breeding colonies. Nest a pile of plant material and sticks, sometimes just bare rock. 1–3 white eggs incubated by both parents for 29–36 days. Fledge at 65–75 days.

OTHER INFORMATION: wingspan up to 360 cm (nearly 12 ft).

CONSERVATION STATUS: not at risk, but threatened by drainage of its habitat, water pollution and human disturbance of breeding colonies.

GREAT FRIGATEBIRD

Fregata minor

Family: Fregatidae

THE MALE GREAT FRIGATEBIRD *develops an inflatable red throat pouch in the breeding season. At the breeding colony, he displays to passing females with wings outstretched and throat pouch inflated.*

DISTRIBUTION: tropical and subtropical Indian and Pacific Oceans, and off Brazil. Breeds on oceanic islands and along coasts.

SIZE: L 86–100 cm (34–39 in); WT 0.64–1.55 kg (1.4–3.4 lb).

FORM: large seabird with long, narrow wings, deeply forked tail and long, hooked bill. Female usually larger than male. Male black with green sheen, pale brown wingbar, scarlet throat. Females mainly dark brown.

DIET: mainly fish, especially flying-fish; also young birds. Feeds on the wing by snatching prey from surface waters, or (flyingfish) from the air. Often bullies other birds in the air to steal their prey.

BREEDING: in colonies, often with other seabirds, whose nests they rob. Single white egg incubated in twig nest by both parents for 40–50 days. Young hatch naked. Fledge at 6 months, but rely on parents for 6 more months.

OTHER INFORMATION: spends most of life at sea, gliding and soaring over coast and ocean. Wingspan over 1.8 m (6 ft), but weight only 1.55 kg (3.4 lb), almost half of it breast muscles.

RED-TAILED TROPICBIRD

Phaethon rubricauda

Family: Phaethontidae

THE RED-TAILED TROPICBIRD'S *long tail feathers are the focus of group courtship displays, that involve an undulating flight in which the streamers wave up and down and from side to side, while the birds call repeatedly.*

DISTRIBUTION: tropical and subtropical west Pacific and Indian Oceans and Indonesia. Breeds on oceanic islands.

SIZE: L 100 cm (39 in); WT 540–750 g (1.2–1.7 lb).

FORM: white seabird with two long, red central tail feathers, not present in juveniles. Bill red, thick and curved, serrated for holding prey. Plumage often tinged pink in breeding season. Pigeon-like flight.

DIET: fish, especially flyingfish; also squid and crustaceans. Catch flyingfish in air by diving on them; otherwise fish from surface or just below.

BREEDING: nest in colonies in holes or on bare ground, under rocks or vegetation. Group displays are thought to synchronize egg-laying in colonies, reducing effects of predation. Often pair for life. Single white egg, blotched reddish-brown, is incubated by both parents for 40–46 days. Chick is left alone while parents forage at sea. Fledges at about 11–15 weeks.

OTHER INFORMATION: usually forages singly or in pairs. Wingspan reaches up to 109 cm (3.6 ft).

CAPE GANNET

Sula capensis

Family: Sulidae

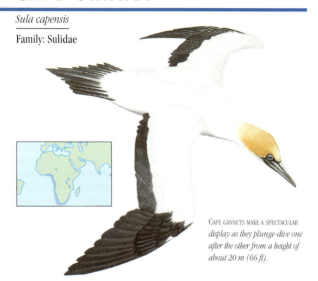

CAPE GANNETS MAKE A SPECTACULAR *display as they plunge-dive one after the other from a height of about 20 m (66 ft).*

DISTRIBUTION: breeds along coasts of South Africa and Namibia. Winters on both coasts of African mainland, north to Gulf of Guinea and Mozambique.

SIZE: L 84–94 cm (33–37 in); WT 2.7 kg (6 lb).

FORM: large black-and-white seabird with yellow head and nape, with a long black stripe down center of throat. Bill powerful and pointed, with fine serrations near tip. Feet large, webbed.

DIET: mainly shoaling fish living in surface waters.

BREEDING: on cliffs or cliff tops, usually on offshore islands, in large colonies. Nest of seaweed and debris. Single blue egg, with chalky white coating, incubated by both parents for 43–44 days. Egg is placed under adults' feet, where it is warmed by the webs, which are rich in blood vessels. To withstand adult's weight, eggs have very thick shells. Chick leaves nest at about 2 months.

OTHER INFORMATION: gregarious, roosting and feeding in large groups. Air sacs under skin of face cushion impact with water when diving.

GREAT CORMORANT

Phalacrocorax carbo

White-breasted cormorant (Africa)
Family: Phalacrocoracidae

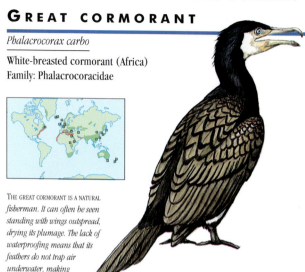

THE GREAT CORMORANT IS A NATURAL
fisherman. It can often be seen
standing with wings outspread,
drying its plumage. The lack of
waterproofing means that its
feathers do not trap air
underwater, making
it easier to dive.

DISTRIBUTION: scattered populations around large stretches of water, both inland and coastal, across eastern North America, Europe, Asia, Africa, Australia and New Zealand.

SIZE: L 80–100 cm (31.5–39 in); WT 1.7–2.7 kg (3.7–5.9 lb).

FORM: aquatic diving bird with long neck, long, thick, hooked bill, short wings, and webbed feet. Color glossy blue-black, with bronze sheen on upper parts, white chin and small yellow throat pouch bordered with white; wing feathers edged in black. In breeding season has white flank patches.

DIET: fish, squid, crabs, tadpoles. Fishes from surface, diving to pursue prey underwater.

BREEDING: spring to summer, in large colonies on rocks and cliff ledges, or in trees inland. Nest a large, rough cup of seaweed and debris, lined with finer material. 3–4 chalky-blue eggs, incubated under webs of parent's feet for 27–29 days. Chicks fledge after 2 months, but still need to be fed by parents for some time.

OTHER INFORMATION: has a steady, flapping flight with occasional glides, often low over water.

GREY HERON

Ardea cinerea

Family: Ardeidae

THE GREY HERON IS A PATIENT *fisherman. It often stands motionless for long periods in wait for prey, or it may stalk a fish or frog cautiously, then seize it with its beak.*

DISTRIBUTION: near shallow fresh and brackish water across much of Europe and Asia, from the British Isles to Japan and south to Java; also found in parts of Africa.

SIZE: L 90–98 cm (35.5–38.5 in); WT 1.02–2.07 kg (2.2–4.6 lb).

FORM: color gray, with white head and neck; black streaks on neck; black eye-stripe extending to black plumes on nape; underwings black; legs brown; bill and eyes yellow; bare skin of face yellowish or green.

DIET: fish, frogs, small mammals; also crustaceans, aquatic insects, mollusks; sometimes young birds.

BREEDING: in spring in temperate regions, almost all year round in the tropics. Breeds in colonies of up to 1,000 birds in trees, bushes or on cliffs, up to 25 m (80 ft) above the ground. Large, untidy stick nest. 3–5 dull, bluish-green eggs, incubated by both parents for 24–25 days. Fledge at 8 weeks.

OTHER INFORMATION: usually flies with its legs trailing and neck drawn well into its body.

HAMMERKOP

Scopus umbretta

Hammerhead
Family: Scopidae

DISTRIBUTION: near fresh or brackish shallow water in south and central Africa and Madagascar.

SIZE: L 50 cm (19.5 in); WT 415–430 g (15 oz).

FORM: color dull brown with dark bill and legs. Has large, backward-pointing crest giving distinct hammer-head shape, and a heavy conical bill.

DIET: mainly fish; also frogs and invertebrates. Feeds mainly at night.

BREEDING: nest a huge domed structure of sticks, grasses, reeds and dead plant stems, up to 2 m (6.6 ft) in diam-

LIKE A HERON, THE HAMMERKOP MAY STALK PREY IN shallow water, stirring up the mud with its feet to disturb prey. It will also snatch food directly from the water while flying low over the surface or hovering above it.

eter and 50 kg (110 lb) in weight, in a tree fork, on large rocks or cliff ledges, or on the ground. A small entrance hole surrounded by mud leads to the nest, which is lined with dry grass and aquatic plants. 3–7 bluish-white eggs, incubated by both parents for about 30 days. Chicks fledge at about 7 weeks.

OTHER INFORMATION: flies like a stork, with head and long neck stretched out in front. Lives alone; sometimes feeds in groups.

WHITE STORK

Ciconia ciconia

Family: Ciconiidae

A WHITE STORK GIVING A CLAPPERING *display, throwing its head back and clappering its mandibles together. This display is used in courtship and when greeting its mate at the nest.*

DISTRIBUTION: breeds on farmland and in marshes of Iberia, northern Africa, central Europe north and east to the Baltic countries and southwest Russia, Asia Minor, south and east Asia. European storks winter in Africa; Asian storks winter in India and south Asia.

SIZE: L 100–115 cm (39–45 in); WT 2.27–4.40 kg (5–9.7 lb).

FORM: a large white stork with long red bill and red legs; back part of wings black both above and below.

DIET: frogs, small fish, worms, snails, beetles and other invertebrates; sometimes eggs and small mammals.

BREEDING: in spring and summer. Builds large, untidy nest of sticks, lined with grasses and twigs, in trees, on cliff ledges and buildings. 3–5 chalky-white eggs, incubated by both parents for 32–38 days. Chicks fledge at 8–9 weeks.

OTHER INFORMATION: flies with head and neck outstretched and legs trailing.

ROSEATE SPOONBILL

Ajaia ajaja

Family: Threskiornithidae

DISTRIBUTION: coasts, lakes and swamps from the southern United States and Caribbean islands south to northern Peru and northern Argentina.

SIZE: L 81 cm (32 in); WT 1.24–1.75 kg (2.7–3.9 lb).

FORM: large pink bird with darker pink wings, bright red on inner leading edge and just above tail, long red legs, bare green skin on head, and a broad, flattened greenish-yellow bill that widens to a spoon-like shape at tip. In breeding season, head may turn buff, and bird has red tuft of feathers on breast.

DIET: mainly small fish, crustaceans, mollusks and other small animals.

THE ROSEATE SPOONBILL FEEDS IN SHALLOW WATER BY *swinging its partly-open bill from side to side. When it touches a small fish, frog or other small prey, its bill snaps shut.*

BREEDING: pair builds untidy nest of twigs, lined with leaves. Nest in groups at tops of trees. 2–5 dull, dirty-white eggs, usually evenly covered with brownish spots and blotches, incubated by both parents for 21–24 days. Chicks fledge at 8 weeks. Sexually mature at about 3 years.

OTHER INFORMATION: courtship involves presenting of twigs.

CONSERVATION STATUS: not at risk. Early in the 20th century numbers were depleted by plume hunters.

GREATER FLAMINGO

Phoenicopterus ruber

Family: Phoenicopteridae

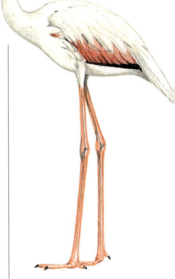

THE GREATER FLAMINGO FEEDS *with its head down, using its strange bill to sieve food from the water. It sucks in water, then uses its tongue to pump it out through comb-like plates, which trap food particles.*

DISTRIBUTION: brackish lakes and coastal lagoons of southern Europe, Africa, southwest Asia, Central America, West Indies and Galapagos Islands.

SIZE: L 125–145 cm (49–57 in); WT 2.1–4.1 kg (4.6–9 lb).

FORM: tall bird with very long neck and legs and down-turned bill. Color washed-out pink, with bright crimson wing-coverts plus black primaries and secondaries; bill pink with black base.

DIET: small fish, crustaceans, mollusks, insects, aquatic plants, grasses.

BREEDING: in very large colonies in shallow water. Each pair build conical nest, usually of mud. 1 or 2 white, speckled eggs, incubated by female for 30–32 days. Chicks join creche after 2–3 days. Independent after 60 days.

MUTE SWAN

Cygnus olor

Family: Anatidae

A FEMALE MUTE SWAN IN A *typical aggressive pose. Swans are powerful birds: a blow from a wing can break a man's arm. The graceful S-shaped curve of the neck is typical of this species.*

DISTRIBUTION: fresh or brackish water lakes and rivers from Britain, southern Scandinavia and the Baltic states to the Black Sea, Asia Minor, and central and eastern Asia. Birds from northern and eastern areas winter farther south and west.

SIZE: L 125–155 cm (49–61 in); WT 7.6–14.3 kg (17–32 lb).

FORM: large waterbird with a long slender neck, pure white plumage, black legs and feet; bill orange with black base and prominent black knob, larger in the male.

DIET: aquatic plants and roots; also worms, shellfish and other types of invertebrate.

BREEDING: pairs for life. Breeds in spring. Large nest of sticks and reeds near water with hollow in center, lined thinly with down. 5–7 bluish-gray or bluish-green eggs incubated by both parents for 34–38 days. Young reared by both parents for 4 months.

OTHER INFORMATION: gregarious outside breeding season.

BAR-HEADED GOOSE

Anser indicus

Family: Anatidae

DISTRIBUTION: breeds in central Asia; feral birds breed in Sweden. Winters in India and northern Burma.

SIZE: L 71–76 cm (28–30 in); WT 2–3.2 kg (4.4–7 lb).

FORM: medium-sized pale gray goose with long neck. Feathers on wings edged white; head white with 2 brownish-black bars over crown and nape; front and nape of neck dark brown, shading to gray on back and breast; sides of neck white; abdomen white; bill yellow with black tip and nostrils; legs yellow to orange; eyes brown.

DIET: mainly plants; some mollusks; also insects when feeding young.

THE BAR-HEADED GOOSE IS A BIRD OF REMOTE *mountain lakes in central Asia. The two horseshoe-shaped blackish bars on its head distinguish it from other geese.*

BREEDING: nest a shallow hollow on raised ground or a rocky ledge, or a twig nest in a tree, thickly lined with down. Usually 4–6 creamy-white eggs, incubated by female for 28–30 days. Young reared by both parents, and stay with them until the start of the next breeding season.

OTHER INFORMATION: call a musical "aang, aang".

MANDARIN DUCK

Aix galericulata

Family: Anatidae

DISTRIBUTION: near fresh water in broadleaved woodlands in far east of Russia, Japan, and northern and eastern China; feral birds breed in England and Scotland.

SIZE: L 47 cm (18.5 in); WT 428–693 g (15–27 oz).

FORM: male has dramatic variegated plumage, with chestnut side-whiskers and wing-fans, and a brown-and-white crest, raised when displaying. Female grayish-brown, with slight crest, blue hindwings and wing-tips, narrow white ring around eye, black bill and yellow legs. Outside breeding season male similar to female except for orange legs and red bill.

A MALE MANDARIN DUCK IN BREEDING PLUMAGE. IN *China and Japan, the Mandarin duck is regarded as a symbol of marital fidelity.*

DIET: mainly acorns and other nuts, seeds, leaves; also insects, snails.

BREEDING: nests in a tree hollow lined with down, 3–15 m (10–50 ft) above ground, often over water. 9–12 creamy-buff eggs, incubated by female for 28–30 days. Young drop from nest within a day of hatching and follow female to water. Reared by female only. Soon independent.

OTHER INFORMATION: a secretive bird that feeds mainly at dusk.

RED-BREASTED MERGANSER

Mergus serrator

Family: Anatidae

THE RED-BREASTED MERGANSER DIVES *for its prey from the water surface, propelling itself with its hind feet. Its bill has serrated edges for gripping slippery fish.*

DISTRIBUTION: breeds along estuaries, coasts and rivers of North America, northern Europe and northern Asia. Winters farther south as far as Gulf of Mexico, Mediterranean and eastern China.

SIZE: L 52–58 cm (20.5–23 in); WT 0.91–1.14 kg (2–2.5 lb).

FORM: long-bodied duck with long neck and long, thin bill with hooked tip. Male has bottle-green head with double crest, white collar, chestnut breast with dark-brown stripes, creamy underparts; dark gray wings, rump and tail, with white wing patch. Legs and bill red. Head, crest and nape of neck of female (and of male outside breeding season) tawny to russet; rest of body gray, with white wingbar.

DIET: fish, shellfish, worms, insects.

BREEDING: May to June. Nests in tree hole lined with plant material, down and feathers. 7–12 (up to 21) creamy-white to greenish-buff eggs, incubated by female for 29–35 days. Chicks fledge at 59 days.

OTHER INFORMATION: feeds by day; spends night on open sea.

COMMON SHELDUCK

Tadorna tadorna

Skelduck, Skeelduck, Shielduck
Family: Anatidae

THE COMMON SHELDUCK MOLTS ITS FLIGHT FEATHERS *after breeding and cannot fly until the new ones grow. Large numbers migrate to favorite bays and estuaries with extensive mudflats for the molt.*

DISTRIBUTION: breeds from northwest and southern Europe east to central Asia, northern China and Siberia. Winters as far south as northwest Africa, Pakistan and southwestern China.

SIZE: L 58–67 cm (23–26 in); WT 0.93–1.45 kg (2–3.2 lb).

FORM: large, goose-like duck with striking plumage of black head, white breast and underparts, with wide chestnut breast band, black belly stripe, chestnut tinge under tail, black and white wings; bill bright red with basal knob; eyes dark.

DIET: feeds by scything its bill through soft mud and sand, or shallow water, for shellfish and other invertebrates, and waterweeds.

BREEDING: breeding groups gather in spring for courtship displays and fights. Nests inland in old rabbit burrows, under bushes or rocks, in hollow trees or buildings; lined with plant material, down and feathers. 8–15 creamy-white eggs incubated by female for 28–30 days, male staying nearby.

OTHER INFORMATION: forms large flocks in winter and during the molt.

MALLARD

Anas platyrhynchos

Family: Anatidae

THE MALLARD IS ONE OF THE COMMONEST DABBLING *ducks. It feeds by dabbling its bill along the water surface, straining food through a comb-like arrangement of plates in the bill. It may also turn tail-up to feed underwater.*

DISTRIBUTION: fresh and coastal waters throughout most of the northern hemisphere outside the tropics, except for the northernmost parts of the tundra. Northern birds migrate south in winter. Introduced to Australasia.

SIZE: L 50–65 cm (20–25.5 in); WT 0.72–1.58 kg (1.6–3.5 lb); male larger than female.

FORM: male (drake) has greenish head, narrow white collar, gray back and underparts, dark purplish-brown breast, orange legs, greenish-yellow bill. Duck is mottled brown with large blue wingpatch edged with white.

DIET: land and water plants, seeds, berries; insects; crustaceans, mollusks and other small invertebrates.

BREEDING: mainly in spring. Drakes promiscuous, and abandon females after mating. Nest a hollow in the ground, usually under bushes or hedges, lined with leaves and grass mixed with down and feathers. 7–12 greenish-gray to tawny-green or bluish eggs, incubated for 28–29 days. Young are led to water soon after hatching. Fledge at 7–8 weeks.

COMMON POCHARD

Aythya ferina

European pochard
Family: Anatidae

THE COMMON POCHARD IS A *diving duck. It dives and swims underwater to feed. As in most diving birds, the legs are set well back on the body for better propulsion in the water.*

DISTRIBUTION: breeds from the British Isles, southern Sweden, southern Finland and eastern Europe east to central Asia and western Siberia. Northern birds migrate to southern regions in winter.

SIZE: L 42–46 cm (16–18 in); WT 0.47–1.24 kg (1–2.7 lb); male larger than female.

FORM: diving duck with short neck and body. Male has chestnut head, finely hatched pale gray back and flanks, black breast and tail, gray legs. Female dark brown with paler cheeks, throat and base of bill; finely hatched on back. Non-breeding male similar to female, but with grayer back.

DIET: waterweeds, grasses, seeds; also small crustaceans, insects and fish.

BREEDING: in spring. Male abandons female after mating. Nest a pile of plant material in vegetation in shallow water, lined with down and feathers. 6–18 pale greenish eggs, incubated for 24–26 days. Young fledge at 7–8 weeks.

OTHER INFORMATION: in flight, wings have gray stripe along trailing edge.

SIBERIAN WHITE CRANE

Grus leucogeranus

Great white crane, White crane
Family: Gruidae

THE ARRIVAL OF THE SIBERIAN *white crane in Russia is a signal that winter is over. The cranes display to each other in pairs and in groups, bowing and "dancing", leaping high into the air.*

DISTRIBUTION: breeds on the tundra in just two areas in northern Siberia. Winters in freshwater wetlands around the Caspian Sea, India and China.

SIZE: L 120–140 cm (47–55 in); WT 4.9–7.4 kg (11–16 lb); male larger than female.

FORM: large pure white bird with black wing-tips, visible only in flight; legs rusty red; face bare, red; bill reddish with slatey-gray tip; eyes yellow.

DIET: bulbs, tubers and corms of underwater plants, shoots and seeds.

BREEDING: in summer. Nest a mound of plant material on the ground. 2 olive-gray eggs, incubated by both parents for 29 days. Fledge at 10 weeks.

OTHER INFORMATION: flight call "koonk, koonk".

CONSERVATION STATUS: highly threatened. A captive breeding program uses eggs taken from wild nests and reared by Common crane foster parents.

AMERICAN PURPLE GALLINULE

Gallinula martinica

Family: Rallidae

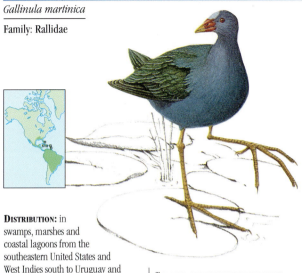

THE AMERICAN PURPLE GALLINULE HAS LARGE FEET *with long, spreading toes for walking and running across the floating leaves of water plants.*

DISTRIBUTION: in swamps, marshes and coastal lagoons from the southeastern United States and West Indies south to Uruguay and northern Argentina. Northern populations migrate south in winter.

SIZE: L 33 cm (13 in); WT 215–257 g (7.6–91 oz).

FORM: color bright purplish-blue, with pale blue forehead shield, conspicuous white undertail coverts; legs yellow; bill red-and-yellow.

DIET: small aquatic invertebrates, seeds and other plant material.

BREEDING: spring. Nest a large cup of grasses and rushes in a clump of reeds, grasses or waterside shrub, or attached to plants in shallow water. 3–12 pale creamy-buff to cream eggs, speckled and finely spotted with dark reddish-brown and light purple, incubated by both parents for 19–22 days. Chicks reared by both parents. Fledge in 6–7 weeks. May breed more than once a year if conditions are right.

OTHER INFORMATION: flight call a cackling "kek kek kek". Flies weakly.

BLACK COOT

Fulica atra

European coot, Coot
Family: Rallidae

A BLACK COOT IN ATTACKING POSTURE. THE LARGE
*flanges on its toes help it to walk across
swampy ground, and provide additional
surface area for propulsion when swimming.*

DIET: waterweeds, grasses, seeds; also
insects, worms, water snails, and other
mollusks.

DISTRIBUTION: freshwater ponds, lakes,
and marshes from the British Isles
and southern Scandinavia south to
northwest Africa and east to Asia and
Australia. Birds from colder parts
migrate to warmer regions in winter.

BREEDING: in spring. Pair builds floating nest of stems and leaves in vegetation along bank of lake or marsh.
5–15 grayish-buff eggs, covered in
dark brown and black speckles and
spots, incubated by both parents for
21–24 days. Chicks brooded by female,
fed by both parents for first month.

SIZE: L 43 cm (17 in); WT 0.56–1.15 kg
(1.2–2.5 lb).

FORM: black waterbird with white bill
and white forehead shield; narrow
white edges to primaries visible in
flight; legs greenish; eyes red.

OTHER INFORMATION: gregarious,
quarrelsome birds, often chasing each
other. Call a "kook" or "kiook".

34

LIMPKIN

Aramus guarauna

Family: Aramidae

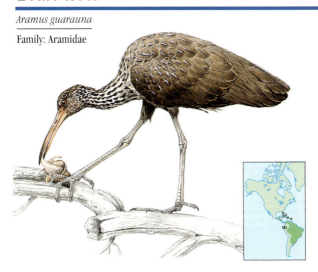

THE LIMPKIN WALKS WITH A STRANGE LIMPING GAIT — *hence its name. It feeds mainly after dark, extracting water snails from their shells and leaving piles of broken shells on the shore.*

DISTRIBUTION: marshes and swamps from Florida and the Caribbean islands south to northern Argentina, excluding areas west of the Andes.

SIZE: L 58–71 cm (23–28 in); WT 0.9–1.27 kg (2–2.8 lb).

FORM: long-legged wading bird, with short, rounded wings, short broad tail, and flattened, slightly down-curved bill. Color dark brown streaked with white; legs grayish-green; bill has dark tip; eyes dark.

DIET: mainly water snails; also other mollusks, reptiles, small animals.

BREEDING: August to October. Pair builds flimsy nest on ground, or in a bush or tree. 4–8 pale creamy-buff or olive-buff eggs, spotted and blotched with various shades of brown and lilac; incubated by both parents. May breed 2 or 3 times a year.

OTHER INFORMATION: call a wailing "krr- oww". Swims well.

35

NORTHERN JACANA

Jacana spinosa

Family: Jacanidae

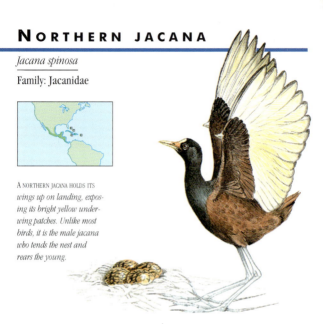

A NORTHERN JACANA HOLDS ITS *wings up on landing, exposing its bright yellow underwing patches. Unlike most birds, it is the male jacana who tends the nest and rears the young.*

DISTRIBUTION: freshwater marshes and lakes of Central America and larger Caribbean islands.

SIZE: L 24 cm (9.5 in); WT 79–112 g (2.8–4 oz).

FORM: color dark brown to chestnut-brown, darker on head, neck and breast, sometimes greenish on head and nape; bill yellow with white band at base and yellow wattle above; legs grayish-green; bill yellow. Male smaller and duller.

DIET: water plants and insects.

BREEDING: wing-raising courtship display. Female mates with several different males. Each male builds nest and rears young. Nest a loose pad of dried waterweeds on a floating leaf. 3–4 golden-buff, buff or golden-brown eggs, patterned with thin black lines, incubated about 22–24 days. Young run and dive soon after hatching, but are guarded by male and brooded under his wings for some time.

OTHER INFORMATION: extremely long toes and claws help spread the bird's weight as it walks and runs over floating leaves. Seldom swims.

GREATER PAINTED-SNIPE

Rostratula benghalensis

Family: Rostratulidae

A FEMALE GREATER *painted-snipe in territorial display. She will court and mate with several males, who will be left to rear the young.*

DISTRIBUTION: scattered populations in swamps, marshes, ponds and rice paddies of Africa, Asia and Australia. Birds from some parts of China migrate south in winter.

SIZE: L 25–28 cm (10–11 in); WT 90–164 g (3.2–5.8 oz); female larger than male.

FORM: has buff crown stripe, white eye patch, white bar on flank, mottled and buff-spotted wings; female brighter, greenish, with chestnut throat and neck, blackish upper breast; male more olive-brown, mottled, with large broad buff V on back, whitish throat, grayish-brown neck and upper breast.

DIET: large insects; crustaceans, mollusks, earthworms; some seeds.

BREEDING: season variable. Spectacular courtship display, initiated by female who flies over her territory, hooting softly. Nest a bed of plant material in marsh vegetation, lined with dry grasses and leaves. 4 pale yellow eggs, heavily marked in black and purple, incubated for about 20 days. Young leave nest soon after hatching.

OTHER INFORMATION: female makes sound as if blowing into empty bottle.

PALEARCTIC OYSTERCATCHER

Haematopus ostralegus

Common oystercatcher, Eurasian oystercatcher
Family: Haematopodidae

THE PALEARCTIC OYSTERCATCHER USES ITS LONG BILL *to open mussels, hammering the hinge of the shell until it opens enough for the bird to insert the tip of its bill and sever the muscle holding it closed.*

DISTRIBUTION: breeds along coasts and on shingle and sand bars along inland rivers, from Europe and Scandinavia east to the Urals and Asia Minor, parts of central Asia, the Kamchatka Peninsula, and around the coasts of the Yellow Sea. Winters farther south to northern Africa, Arabia, India and southeast China.

SIZE: L 32 cm (17 in); WT 430–675 g (15.2–23.8 oz).

FORM: large shore bird with very striking black-and-white plumage, long orange bill and pink legs; throat black in summer, white in winter; white wingbar visible in flight; eyes red, with distinct orange rim.

DIET: mussels, cockles, crustaceans, insects and other invertebrates.

BREEDING: in spring. Makes scrape on shingle, sand, among rocks or in vegetation, sometimes lined with dead plant material or small stones. 2–4 pale buff eggs, spotted and blotched with black, incubated by both parents for 24–27 days. Chicks leave nest at 1–2 days, and are fed and guarded by both parents. Fledge at about 5 weeks.

OTHER INFORMATION: noisy and gregarious. Groups take part in "piping ceremony", walking to and fro, facing ground, making plaintive calls.

EUROPEAN GOLDEN PLOVER

Pluvialis apricaria

Family: Charadriidae

THE GOLDEN PLOVER'S *mottled plumage camouflages it as it incubates its eggs. If danger threatens, it may lure predators away.*

DISTRIBUTION: breeds on moorlands, heaths and tundra in northern Europe, Scandinavia, northern England, Scotland, northwest Russia, Iceland and Greenland. Winters in the British Isles, western Europe, North Africa and the Mediterranean.

SIZE: L 28 cm (11 in); WT 165–260 g (5.8–9.2 oz).

FORM: has rounded head, short neck, pointed wings, slender medium-length bill. Color dark brown flecked with golden-yellow; tail and rump darker; underwings white. In summer has black face, throat and underparts, which in northern birds are bordered in white.

DIET: insects; mollusks; some plants.

BREEDING: in spring. Nest a shallow scrape lined with grasses and lichens. 3–4 buff, sandy or yellowish eggs, sometimes tinted reddish or olive, blotched with blackish-brown, incubated by both parents for 27–28 days. Young reared by both parents; independent in about a month.

OTHER INFORMATION: forms large flocks in winter.

WESTERN CURLEW

Numenius arquata

Eurasian curlew, Curlew
Family: Scolopacidae

THE WESTERN CURLEW USES ITS LONG BEAK TO PROBE *deep into mud or marshy ground for invertebrates such as crabs, snails and worms. Its large feet help prevent it sinking into the mud.*

DISTRIBUTION: breeds on moorland and other damp open country, marshes, forest clearings and coastal dunes in central and northern Europe and Asia. Winters along muddy shores and estuaries farther south.

SIZE: L 50–60 cm (19.5–23.5 in); WT 540–1,040 g (19–36.7 oz).

FORM: a large wading bird with a long, downward-curved bill. Color light brown to chestnut with darker streaks, and finer markings on head; rump white; legs grayish-green; bill dark, reddish. Paler in winter.

DIET: crustaceans, mollusks and other marine invertebrates; fish, earthworms; some plants; also insects, berries, seeds.

BREEDING: in spring. Nest a large scrape lined with plant material. 3–5 light green to olive eggs, spotted, speckled and blotched with dark olive, or dark purplish- or reddish-brown, especially at larger end, incubated by both parents for 26–30 days.

OTHER INFORMATION: flies fast, with slow wing-beats.

COMMON SNIPE

Gallinago gallinago

Family: Scolopacidae

A COMMON SNIPE TAKES TO THE AIR IN A *territorial display. As it dives toward the ground, it spreads out its outer tail feathers, to produce a vibrating, drumming sound.*

DISTRIBUTION: breeds in marshes and damp open country throughout northern and central North America, Europe, and Asia. Winters near fresh water farther south to northernmost South America, north and central Africa, the Middle East and southeast Asia.

SIZE: L 25–27 cm (10–10.5 in); WT 116–128 g (4.1–4.5 oz).

FORM: wading bird with very long, straight bill and strongly striped plumage. Color tawny brown with dark brown, black and russet markings; flanks cream with dark bars; belly white; crown black with tawny central stripe; sides of head brown with tawny

stripes above and below eye; legs pale green; bill blackish; eyes dark.

DIET: earthworms, insects, mollusks, crustaceans; some plants.

BREEDING: in spring. Nest a shallow scrape lined with grass. 3–4 pale green, olive or buff eggs, marked with shades of brown, olive and purplish-gray, incubated by female alone for 18–20 days. Young leave nest almost immediately and are reared by both parents. Fledge at 19–20 days.

OTHER INFORMATION: often forms small groups, which may perform aerobatics together. Active mainly at dusk.

STONE CURLEW

Burhinus oedicnemus

Thick-knee
Family: Burhinidae

DISTRIBUTION: breeds in open country with short vegetation, bare sand or shingle, or in scrub from western Europe to Central Asia, India, IndoChina and North Africa. Some birds winter farther south.

SIZE: L 40–44 cm (16–17 in); WT 290–535 g (10.2–18.9 oz).

FORM: color light brown, flecked and streaked with darker brown; white wingbar bordered with black, visible when standing; two white wingbars visible when flying; wing has dark leading edge, very marked when flying; belly and flanks white; legs yellow; bill short and straight, with black tip and yellow base; eyes yellow, staring.

DIET: earthworms, mollusks, insects; sometimes mice, young birds, lizards.

A STONE CURLEW PERFORMS A GREETING DISPLAY TO *its mate. Stone curlews are sometimes called thick-knees because of their knobbly leg joints.*

BREEDING: in spring. Nest a shallow scrape on the ground, unlined or thinly lined with plant fragments, small stones or rabbit dung. 2–3 whitish to pale brown eggs, mottled and streaked with brown and gray, incubated by both parents for 25–27 days.

OTHER INFORMATION: flight call "coo-eek". Runs in short steps, stopping and standing upright at intervals, bobbing head down, tail up.

EGYPTIAN PLOVER

Pluvianus aegyptius

Family: Glareolidae

AN EGYPTIAN PLOVER PERFORMS *a wing-raising display to greet its mate, showing off its beautiful wing markings. This species is said to enter the mouths of crocodiles to feed on food lodged between their teeth.*

DISTRIBUTION: along river banks and lakesides of tropical Africa, from Senegal and Ethiopia south to Uganda and Angola.

SIZE: L 19–21 cm (7.5–8 in); WT 73.4–90.5 g (2.6–3.2 oz).

FORM: bold plumage pattern, with black crown and nape, black band passing through eye and over shoulders and back, joining greenish-black breast band; rest of head and throat white; breast and belly chestnut-brown to orange; lower back and tail slatey-gray; flight feathers in black and white pattern on both sides; legs gray; eyes dark.

DIET: insects, small mollusks and other invertebrates

BREEDING: season variable. 2–3 cream eggs with dark brown and black speckles and scribbles, partly or completely buried in sand and incubated by both parents, who sit on mound of sand. Chicks also buried in sand, kept moist with water brought by parents.

OTHER INFORMATION: catches insects as it runs along ground, or leaps into the air after them.

PIED AVOCET

Recurvirostra avosetta

Avocet
Family: Recurvirostridae

A PIED AVOCET DEFENDS ITS NEST AND EGGS AGAINST *a predator. When feeding, it sweeps its curved bill from side to side in shallow water, detecting small prey by touch.*

DISTRIBUTION: breeds in freshwater or brackish wetlands, both coastal and inland, in locations across Eurasia from southeast England to the Far East. Most populations winter farther south.

SIZE: L 42–45 cm (16.5–18 in); WT 228–435 g (8–15.4 oz).

FORM: long-legged wader with a slender upcurved bill and webbed feet. Color mainly white; black on crown and nape extends to bill and just below eye; black bands on wings and wingtips; legs bluish-gray; bill black; eyes brown.

DIET: crustaceans, mollusks; insects.

BREEDING: in spring. Nest a shallow scrape on bare ground or in short vegetation near water's edge, unlined or thinly lined with dead plant material. 3–5 pale brownish-buff eggs, marked with small spots and blotches of black or gray, incubated by both parents for 22–24 days. Young leave nest soon after hatching.

OTHER INFORMATION: live in large groups of up to 200 birds.

LONG-TAILED SKUA

Stercorarius longicaudus

Family: Stercorariidae

DISTRIBUTION: breeds on tundra throughout the high Arctic. Winters in the southern Pacific and southern Atlantic Oceans.

SIZE: L 50–58 cm (19.5–23 in); WT 236–358 g (8.3–12.6 oz).

FORM: a small, slightly-built seabird with long narrow wings and long central tail feathers, which may extend up to 18 cm (7 in) beyond rest of tail. Color dark brown above, white below, with blackish-brown crown, wing tips and tail, and pale yellow sides of neck; legs slatey-gray to black; bill black; eyes dark brown.

DIET: fish and marine invertebrates; feeds on lemmings and insects in

THE LONG-TAILED SKUA IS A PIRATE, CHASING AND *harassing other seabirds until they disgorge the food they are carrying. In the breeding season it feeds mainly on lemmings.*

breeding season. Uses piracy less often than other skuas.

BREEDING: in summer. Breed in loose colonies. Nest a shallow scrape, unlined or thinly lined with plant material, in peat or moss. 1–3 olive-green to olive-brown, pale green or buff eggs, spotted, blotched or scrawled with dark brown or pale gray, especially around larger end, incubated by both parents for about 23 days. Young leave nest after 2 days. Cared for by both parents.

OTHER INFORMATION: graceful soaring flight; also hovers like a tern.

LESSER NODDY

Anous tenuirostris

Lesser noddy tern
Family: Sternidae

DISTRIBUTION: offshore and inshore waters around certain islands in the Indian Ocean; roosts on land at night; breeds in the Seychelles, and off Western Australia, on the Abrolhos Islands, Pelstart and Wooded Islands. Remain nearby for rest of year.

SIZE: L 28–30 cm (11–12 in).

FORM: small tern with long, quite slender bill and squarish or slightly forked tail. Color dark grayish-brown; pale gray to whitish on forehead and crown, shading to blackish on lower face; black mark occurs around eye; legs and bill gray.

DIET: mainly fish and squid. Plunge-dives for fish.

THE LESSER NODDY HUNTS FOR FISH AND SQUID FAR *out to sea. Noddy terns are named for their greeting displays, which involve much head-nodding.*

BREEDING: in spring. Nest a platform of seaweed and leaves bound together and fixed to tree branch with guano, with a slight depression on top for egg. Single pale creamy-white egg, spotted and blotched with reddish-brown or purplish-gray, incubated by both parents for about 32–35 days. Fledges at about 6 weeks.

OTHER INFORMATION: very gregarious, living in large flocks. Call loud and rattling; also croaks and purrs.

CONSERVATION STATUS: locally abundant but of very restricted distribution.

ARCTIC TERN

Sterna paradisaea

Family: Sternidae

THE ARCTIC TERN MAKES A *round trip of some 32,000 km (20,000 mi) every year, as it migrates from its breeding grounds in the Arctic.*

DISTRIBUTION: breeds on marshy tundra and saltmarshes throughout the Arctic and sub-Arctic; winters farther south, as far as the Southern Ocean around the Antarctic pack ice.

SIZE: L 33–38 cm (13–15 in); WT 86–127 g (3–4.5 oz).

FORM: color medium gray above, with paler underparts; underwing has narrow black band on trailing edge; crown and nape black; legs and bill deep red; bill black outside breeding season. Has round head and long tail.

DIET: fish and crustaceans.

BREEDING: in northern summer. Forms large breeding colonies, sometimes with other seabirds. Nest a shallow scrape made by female, on ground or in moss, unlined or thinly lined with plant material, small pebbles or shells. 1–3 bluish-white, creamy, buff, pale greenish, olive or deep brown eggs, spotted, speckled, blotched and scribbled with blackish-brown, black or dark olive. Incubated by both parents for 20–22 days. Young leave nest soon after hatching, and are cared for by both parents for about 3 weeks.

OTHER INFORMATION: call a rasping "tr-tee-ar".

SOOTY TERN

Sterna fuscata

Wideawake
Family: Sternidae

DISTRIBUTION: all tropical and sub-tropical oceans; breeds on islands. Sometimes wanders as far north as Maine and western Europe.

SIZE: L 43–45 cm (17–18 in); WT 147–220 g (5.2–7.8 oz).

FORM: a large tern, with deeply-forked tail. Color brownish-black above, white below, with white forehead; tail edged with white; legs and bill black; eyes brown.

DIET: fish, especially flyingfish, and squid. Huge flocks may gather where schools of predatory tuna drive smaller fish to the surface.

THE SOOTY TERN SPENDS ALMOST ALL ITS LIFE IN THE *air. It even feeds on the wing, hovering, then swooping to snatch small fish and squid from the water surface.*

BREEDING: season variable – breeds at less than annual intervals. Forms very large, noisy densely-packed breeding colonies. Pairs defend territory by fighting neighbors. Nest a shallow scrape in sand, sometimes surrounded by leaves. 1–2 whitish or pinkish-white eggs, spotted and blotched with brown or violet, incubated by both parents for 24–26 days. Fledge at about 6 weeks.

OTHER INFORMATION: sometimes carried long distances inland by tropical storms. Call a high "wacky-wack".

HERRING GULL

Larus argentatus

Family: Laridae

A YOUNG HERRING *gull pecks at the red spot on its mother's bill to persuade her to regurgitate some food. Baby Herring gulls will stab at anything red.*

DISTRIBUTION: breeds on cliff ledges, buildings, shingle, dunes and islands throughout the northern hemisphere. Winters along coasts and around lakes and reservoirs. Some birds wander south toward the subtropics in winter.

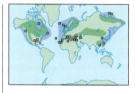

SIZE: L 56–66 cm (22–26 in); WT 0.72 –1.39 kg (1.6–3.1 lb).

FORM: adult birds vary greatly in size and color. Wings and back dark to silvery-gray, with white underparts; head white, streaked with brown in winter; legs and feet pink, yellow in some races; bill yellow with red spot near tip.

DIET: mainly fish, crustaceans and mollusks; also other birds, eggs and young. Scavenges on refuse and carrion.

BREEDING: in spring and summer. Nests in colonies. Pair build large nest of seaweed, grass and debris in a scrape or hollow. 2–3 pale olive, buff, green or brown eggs, spotted, speckled and blotched with black, blackish-brown or dark olive. Incubated by both parents, especially female, for 25–33 days. Young cared for by both parents. Fledge at about 6 weeks.

OTHER INFORMATION: open shellfish by dropping them onto rocks.

BLACK-HEADED GULL

Larus ridibundus

Common black-headed gull
Family: Laridae

DISTRIBUTION: breeds on marshes, shingle, dunes and islands from Iceland, British Isles and Spain across Eurasia to Kamchatka Peninsula and Japan. Winters along coasts and lake shores west to northeast North America, south to central Africa, India and Malaysia.

SIZE: L 41 cm (16 in); WT 195–327 g (6.9–11.5 oz).

FORM: color pale gray above, white below, with black wingtips; hood dark brown; legs and bill red; eyes brown. In winter lacks dark hood, but has dark spot behind eye.

DIET: fish, crustaceans, mollusks, worms; insects.

THE CONTRAST BETWEEN BLACK-HEADED GULL'S DARK *hood and white neck is displayed to advantage during courtship. Pairs stand alongside each other, then "headflag", sharply turning their heads toward and away from each other.*

BREEDING: nests in shallow scrape or on low vegetation; nest of plant material built mainly by male. 2–6 greenish, olive, buffish or brown eggs, spotted, blotched and scribbled in black, brown, or olive, especially around larger end; often also with gray markings. Incubated by both parents for 22–27 days. Young leave nest soon after hatching; are cared for by both parents. Fledge at about 5–6 weeks.

OTHER INFORMATION: plucks prey from surface of water or wades in the shallows; catches insects in the air.

BLACK SKIMMER

Rynchops nigra

Family: Rynchopidae

DISTRIBUTION: inshore waters along coasts, estuaries, lagoons and other bodies of water, from New Jersey in the United States south to Chile.

SIZE: L 40–50 cm (16–19.5 in); WT 212–392 g (7.5–13.8 oz); male larger than female.

FORM: long-winged seabird with short legs and heavy head; the thick bill is flattened sideways, with the lower mandible longer than the upper one. Color black, with white face and underparts; legs red; bill red with black tip.

DIET: mainly small fish.

BREEDING: in spring and summer. Forms compact breeding colonies,

THE BLACK SKIMMER HAS AN EXTRA-LONG LOWER *mandible. It feeds while flying low over the water, "plowing" the surface with its lower mandible, which snaps shut the moment it touches a fish.*

often near other seabirds. Nest an unlined scrape in sand or shell debris above high tide mark. 3–6 whitish or buffish eggs, heavily spotted, blotched and scribbled with dark brown, purple and gray, incubated by both parents, especially female, for 21–23 days. Cared for by both parents.

OTHER INFORMATION: young hatch with mandibles of equal length, and can pick up food dropped on ground in front of them, which their parents are unable to do. Feeds mainly at dusk and dawn, and in moonlight.

ATLANTIC PUFFIN

Fratercula arctica

Common puffin
Family: Alcidae

ATLANTIC PUFFINS GREET EACH OTHER *at the nest hole. They can carry up to 10 small fish at a time, held in place between the tongue and the upper mandible, leaving the lower mandible free to catch more.*

DISTRIBUTION: breeds on cliff-tops around coasts and islands of North Atlantic, from northeast North America and Greenland to Svalbard and Novaya Zemlya. Winters farther south.

SIZE: L 28–30 cm (11–12 in); WT 381 g (13.4 oz).

FORM: stocky bird with large head and short wings; legs short set well back on body, giving it an upright stance on land. Breeding plumage black on head, back, wings and tail; underparts white; large white cheek patches; bill very large, flattened sideways, striped with red, yellow and gray or blue; legs orange-red; eyes yellow with red eye-ring. In winter, bill smaller and dull yellow, face dusky.

DIET: small fish, mollusks and other marine invertebrates, caught by swimming underwater.

BREEDING: in summer. Atlantic puffins breed in large colonies where they excavate burrows or use old rabbit or shearwater burrows, natural hollow or rock crevice. Nest a shallow scrape deep inside tunnel or crevice, scattered with plant material. 1 (rarely 2) white egg, often with faint brown or purplish blotches, incubated by both parents, mainly by female, for 40–43 days. Chick cared for by both parents, but deserted in 40 days.

BLACK GUILLEMOT

Cepphus grylle

Sea pigeon
Family: Alcidae

DISTRIBUTION: scattered populations across temperate, sub-Arctic, and Arctic regions of northern hemisphere. Breeds on rocky coasts and sea inlets; winters mainly offshore.

SIZE: L 30–36 cm (12–14 in); WT 405 g (1 lb).

FORM: largish seabird with large head, long bill and short tail; legs short set well back on body, giving upright stance on land. Breeding adult black, with large white wingpatch. Winter plumage white with black-mottled upperparts and crown, underparts gray-mottled; wingpatch less distinct.

DIET: bottom-dwelling fish, marine invertebrates, seaweed; plankton.

BREEDING: in summer, in a rock crevice, near base of cliff, in colonies.

A BLACK GUILLEMOT WITH ITS PREY. THIS BIRD "FLIES" *underwater in search of bottom-living fish, using its wings like flippers. At the surface it propels itself along with its large, webbed feet.*

Nest a scrape, unlined or lined with debris. 1–3 whitish, pale buff or bluish-green eggs, spotted and blotched with black, pale gray or reddish-brown, incubated by both parents (female by day, male by night) for 21–25 days. Young helpless at first. Cared for by both parents. Fledge at 34–40 days.

OTHER INFORMATION: flies fast and low over surface of sea.

ANDEAN CONDOR

Vultur gryphus

Family: Cathartidae

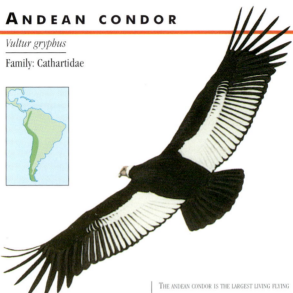

THE ANDEAN CONDOR IS THE LARGEST LIVING FLYING *bird. Its wingspan of over 3 m (10 ft) enables it to glide great distances across the Andes.*

DISTRIBUTION: the Andes mountains of South America, from western Venezuela to Tierra del Fuego.

SIZE: L 110 cm (43 in); WT 9.6–13.6 kg (21–30 lb).

FORM: very large vulture; wings long with squarish tips; tail broad. Color glossy black; wing coverts silvery-gray; white ruff around neck, not quite meeting in front; head and neck bare, reddish; male has fleshy dark red or black comb; legs gray; bill white; eyes brown.

DIET: carrion, refuse.

BREEDING: in spring, in mountain cave or cleft of rock. Single (rarely 2) white egg incubated by both parents for 54–58 days. Chick reared by both parents. Fledges at 16 months. Breeds every other year.

OTHER INFORMATION: uses excellent eyesight to detect carrion from a great height. Also watches other vultures spiraling down to carcasses.

CONSERVATION STATUS: not at risk, but numbers are declining as a result of human disturbance.

LAMMERGEIER

Gypaetus barbatus

Bearded vulture
Family: Accipitridae

THE LAMMERGEIER IS FAMOUS
*for its habit of smashing
bones (and tortoise shells)
by dropping them onto
rocks from a great height.
Its specialized tongue
acts as a scoop to
extract the bone
marrow.*

DISTRIBUTION: mountain areas from southern Europe and Africa to the Middle East, central Asia and Tibet.

SIZE: L 110–115 cm (39–45 in); WT 5.0–6.75 kg (11–15 lb); female larger than male.

FORM: a large vulture with a "beard" of bristly black feathers; wings long, narrow, pointed; tail long, diamond-shaped. Color blackish-gray back, wings and tail; crown and face dirty white, with black band through eye; underparts and leg feathers tawny; legs fully feathered; feet dark gray; bill black.

DIET: mostly carrion.

BREEDING: in late winter or early spring. Nest large, shallow, on a mountain ledge; made of twigs and branches, lined with animal bones. 1–2 whitish eggs, blotched with brown or purple, incubated by female for 53 days. Usually only one chick is reared by both parents. Fledges at 107–117 days.

OTHER INFORMATION: territorial; lives in pairs; uses nest all year.

CONSERVATION STATUS: not at risk. Becoming rare in parts of its range.

ASIAN WHITE-BACKED VULTURE

Gyps bengalensis

Oriental white-backed vulture,
Indian white-backed vulture
Family: Accipitridae

THE ORIENTAL WHITE-BACKED VULTURE GETS UP *after the sun has warmed the land and set up spiraling thermal currents. It circles in the sky as it searches for carcasses, then spirals down to compete with other vultures.*

DISTRIBUTION: wooded and cultivated areas in the Indian subcontinent, Burma and IndoChina.

SIZE: L 90 cm (35 in); WT 3.5–6 kg (7.7–13.2 lb).

FORM: a large vulture, with partially feathered legs. Color of upperparts slatey-gray to blackish or brown, with paler secondaries; underparts blackish-brown, with white rump; has grayish-buff ruff; legs grayish to black; bill dark; eyes yellowish-brown. In flight conspicuous white primary wing coverts are visible.

DIET: carrion, especially dead cattle.

BREEDING: in spring; forms breeding colonies 12–18 m (40–60 ft) above ground in trees along rivers or near villages. Nest made of sticks, lined with leaves, dung and pieces of skin. Single white egg, faintly marked with red, brown, gray or lavender, incubated by both parents for 45–52 days. Chicks reared for about 3 months.

OTHER INFORMATION: wingspan 210 cm (almost 7 ft). Makes range of whistles, hisses, squeals and grunts.

SECRETARY BIRD

Sagittarius serpentarius

Family: Sagittariidae

DISTRIBUTION: grasslands, savannas, open country and cultivated land of Africa south of the Sahara.

SIZE: L 125–150 cm (49–59 in); WT 3.41–3.81 kg (7.5–8.4 lb).

FORM: a tall, slender, very long-legged bird with long wings, a long tail with long central feathers; small feet. Color bluish-gray with grayish-white underparts; tail gray, black-and-white; bare skin of face bright pink; crest of long, loose black feathers; legs flesh-colored; bill dark gray; eyes brown.

DIET: small mammals; eggs and young birds; lizards, snakes.

THE SECRETARY BIRD WAS SAID TO RESEMBLE AN *old-fashioned secretary with a quill pen stuck behind his ears. It is famed for its technique of killing snakes, stamping to disturb its prey, then seizing it with its small feet.*

BREEDING: starts May or June. Nests up to 12 m (40 ft) above ground in tree or on cliff. Nest a platform of sticks up to 2.4 m (8 ft) across. 2–3 whitish, pinkish or buff eggs, streaked with reddish-brown, incubated by female for about 45 days. Chicks fledge at 65–80 days.

OTHER INFORMATION: hunts by striding along the ground; seldom flies. Has spectacular courtship display in which pair soar high into sky, trailing their legs and long tails, and growling.

OSPREY

Pandion haliaetus

Family: Pandionidae

The osprey hovers above the water some 30 m (100 ft) in the air, then swoops down and seizes fish with its feet. Long claws and tubercles on its feet help it grip the slippery prey. The bird returns to a favorite perch to devour its meal.

DISTRIBUTION: breeds on cliffs, in trees or on ground, near fresh water or along coasts, across northern hemisphere from sub-Arctic to subtropics; also in Australia and adjacent parts of southeast Asia. Some birds migrate to warmer areas in winter.

SIZE: L 55–58 cm (21.5–23 in); WT 1.22–1.90 kg (2.7–4.2 lb); female larger than male.

FORM: large eagle, with long wings. Color dark brown above, white below, with white head, and dark eye-stripe; dark wrist patches on underwings; female has streaked band on breast.

DIET: mainly fish; also small mammals and insects.

BREEDING: in spring. Large, untidy nest of sticks, seaweed, bones and debris, added to year by year. 2–4 creamy or yellowish eggs, spotted and blotched with dark brown or chestnut-red. Incubated by both parents, especially female, for 35–38 days. Cared for by both parents. Fledge at 51–59 days.

OTHER INFORMATION: wings angled at wrist when in flight.

SNAIL KITE

Rostrhamus sociabilis

Everglade kite
Family: Accipitridae

THE SNAIL KITE'S BILL IS HIGHLY adapted for its specialized diet of apple snails. The long sharply curving upper mandible is used to hook the snails from their shells.

DISTRIBUTION: freshwater marshes with open water in Florida and Cuba, and from east Mexico south to Argentina. South American birds migrate north; North American birds migrate south in winter.

SIZE: L 43 cm (17 in); WT 378 g (13.3 oz).

FORM: smallish bird of prey with large, broad wings. Color of male black, with broad white band at base of tail; legs red. Female buffish, heavily streaked, white stripe over eye, black tail with white band.

DIET: aquatic apple snails.

BREEDING: starts February. Forms breeding colonies in plants growing in water. Male builds nest, a platform of twigs lined with finer plant material. 2–4 white eggs, speckled and spotted with various shades of brown or reddish-brown, with large blotches of same color. Fledge at about 4 weeks.

OTHER INFORMATION: has slow, flapping flight as it scans water for snails.

CONSERVATION STATUS: not at risk. The Florida population came close to extinction, but is now increasing.

RED KITE

Milvus milvus

Family: Accipitridae

THE RED KITE IS A GRACEFUL FLIER *with a deeply-forked tail and long, narrow wings. It flies slowly, gliding while it searches for prey, then twisting down to swoop on a rabbit, mouse or other small creature.*

DISTRIBUTION: open country, hills, and cultivated land from Wales and western Europe to central Europe, Asia Minor and northwest Africa. Northern populations winter farther south in the Mediterranean and Africa.

SIZE: L 61 cm (24 in); WT 0.8–1.3 kg (1.8–2.9 lb).

FORM: color dark brown above, with pale borders to feathers; underparts reddish-brown to rusty-brown with darker streaks; more tawny on breast; head gray with dark streaks, darker gray around eyes; legs, cere and base of bill yellow; rest of bill black.

DIET: small mammals, birds, frogs, fish, insects and earthworms.

BREEDING: in spring, in wooded areas near open country. Builds nest of sticks and debris up to 30 m (100 ft) above ground in tree. Male brings material and female builds. Nest is added to annually. 1–5 white eggs, spotted or blotched with reddish- or purplish-brown, incubated mainly by female for 28–30 days. Fledge at 45–50 days. Return to nest for further 2 weeks.

OTHER INFORMATION: call a high-pitched "wee-oo-wee-oo-wee-oo".

CONSERVATION STATUS: declining in range and numbers.

PALE CHANTING-GOSHAWK

Melierax canorus

Gray chanting-goshawk
Family: Accipitridae

THE PALE CHANTING-GOSHAWK GETS *its name from the male's melodious whistling "chants", which he makes in the breeding season from a song perch high in a tree, spreading his tail as part of his display.*

DISTRIBUTION: southern and eastern Africa, in dry savanna, thorn scrub, steppe and desert with scattered trees.

SIZE: L 42.5–45 cm (16.7–17.7 in); WT 685 g (1.5 lb); female larger than male.

FORM: large gray bird (darker above than below), with pink or reddish legs; pink to orange cere; primaries black; tail barred black-and-white, with gray central feathers; underparts and leg feathers finely barred gray-and-white; bill dark; eyes red.

DIET: mainly insects and lizards; also small mammals and birds.

BREEDING: in summer. Nest a platform of sticks high in a tree or on a telegraph pole, lined with fur, dung and debris. 1–2 pale bluish or greenish-white eggs, incubated by female for about 35 days. Chicks fledge at 44 days.

OTHER INFORMATION: a versatile hunter, swooping down on prey from a high look-out post, stalking it along the ground or snatching it in mid-air.

NORTHERN SPARROWHAWK

Accipiter nisus

Sparrowhawk,
Family: Accipitridae

THE SPARROWHAWK HAS SHORT WINGS *and a long tail that are adaptations to the twisting and turning flight needed to pursue prey among the trees.*

DISTRIBUTION: breeds in Europe, Asia and northwestern tip of Africa; northern birds migrate south in winter as far as Persian Gulf and southeast Asia. Favors wooded areas, parks, gardens.

SIZE: L 28–38 cm (11–15 in); WT 150–325 g (5.3–11.5 oz).

FORM: color of male slatey-gray above, tail brownish; wings with dark bars; underparts finely barred reddish-brown; cheeks rufous; fleshy area at base of bill; eyes yellow; bill dark. Female browner, with whitish underparts, reddish-brown flanks and white stripe above eye.

DIET: mainly small birds caught on the wing; also insects and small mammals, such as voles and mice.

BREEDING: in spring. Nest built mainly by female, in a tree, a loose platform of twigs lined with leafy twigs. 2–7 bluish-white eggs, spotted, blotched or streaked with dark brown, incubated by female for 32–35 days. Male brings food. Female broods chicks at first, then both parents feed them. Chicks fledge at 32 days, but stay with parents for further month.

OTHER INFORMATION: has low, fast flap-and-glide flight.

BALD EAGLE

Haliaeetus leucocephalus

Family: Accipitridae

THE BALD EAGLE'S POWERFUL TALONS *can snatch slippery fish from the water. It is a faithful mate, reinforcing its devotion by spectacular aerial mating displays, in which both birds lock talons and tumble through the sky.*

DISTRIBUTION: breeds along sea coasts, large rivers and lakes across much of North America; northern populations winter farther south.

SIZE: L 79–94 cm (31–37 in); WT 3.63–6.40 kg (8–14 lb).

FORM: a large dark brown eagle with white head and tail, very large yellow bill, yellow legs and eyes.

DIET: mainly fish; also small mammals, birds and carrion.

BREEDING: in late fall in south, spring in north. Pair for life. Builds large, deep nest of branches and sticks, lined with grass and other plants, in tall tree, on cliff or on small island. Nest is added to annually. 4–12 bluish-white eggs, incubated by female for 29–39 days. Cared for by female at first, male bringing food; then both parents bring food. Chicks fledge at about 37 days.

OTHER INFORMATION: gathers in large numbers where salmon spawn.

CONSERVATION STATUS: not at risk, but numbers much reduced by hunters, egg collectors and pesticides.

BATELEUR EAGLE

Terathopius ecaudatus

Family: Accipitridae

THE BATELEUR EAGLE IS ADEPT AT *hunting both snakes on the ground and birds in the air. It soars rapidly across the savanna, scouring the air and ground below for prey.*

DISTRIBUTION: savannas, thorn scrub and open country in Africa, south of the Sahara. Roosts in trees.

SIZE: L 56–61 cm (22–24 in); WT 1.93–2.95 (4.3–6.5 lb).

FORM: a stocky eagle with long tapering wings, especially in male; almost no tail. Color of male black with chestnut (rarely white) back, tail and undertail coverts; upper wing tawny; legs, cere and eyes red; bill red, shading to yellow with dark tip. Female has more extensively tawny wings.

DIET: mainly birds; also small mammals, lizards, snakes, some invertebrates, carrion. Often force other predators to give up their catches.

BREEDING: builds nest of sticks lined with plant material up to 18 m (60 ft) above ground in tree. Single chalky white egg, incubated by female for 42–43 days. Young cared for by both parents. Fledges at 90–125 days.

OTHER INFORMATION: solitary. Spends most of time soaring in search of prey.

GOLDEN EAGLE

Aquila chrysaetos

Family: Accipitridae

A GOLDEN EAGLE LUNGES SIDEWAYS *at an opponent during a territorial dispute. A skilled flier, it can reach speeds of up to 180 km (120 mi) per hour as it sweeps on prey or displays with its mate.*

DISTRIBUTION: breeds throughout much of North America, Europe, North Africa, the Middle East and Asia, outside the tropics. Northern populations migrate south in winter.

SIZE: L 76–99 cm (30–39 in); WT 3.48–4.91 kg (7.7–10.8 lb).

FORM: a large, powerful brown eagle with a golden/tawny tinge over crown and nape; tail faintly barred; legs and cere yellow; bill large, yellowish-black; eyes brown; tail square; legs feathered.

DIET: mainly small mammals to size of small fawns; also birds and carrion.

BREEDING: in late winter in south, spring to summer in north. Builds large nest of branches and twigs, lined with leafy twigs, grasses or conifer needles, in tree or on cliff or high rock. Adds to nest annually. 1–3 white eggs, often spotted or blotched with brown, chestnut and pale gray, incubated usually by female for 43–45 days. Brooded by female at first, then fed by both parents. Fledge at 63–70 days.

OTHER INFORMATION: pairs for life. Uses nest all year.

CONSERVATION STATUS: not at risk, but numbers have been reduced.

WESTERN HONEY-BUZZARD

Pernis apivorus

Family: Accipitridae

DISTRIBUTION: breeds from southern England and western Europe, east to central Asia and western Siberia; winters in Africa south of the Sahara. Found in mature woodland.

SIZE: L 52–60 cm (20.5–23.5 in); WT 0.51–1.05 kg (1.1–2.3 lb).

FORM: dark brown bird of prey; underparts brown or white, barred with brown; double dark bar near base of tail and black bar at its tip; crown and sides of head gray; legs, lower part of cere and eyes yellow; rest of cere and bill blackish.

DIET: mainly bees, wasps and hornets, their larvae, pupae and sometimes nests. Also other insects, earthworms, small vertebrates; berries.

THE WESTERN-HONEY BUZZARD FEEDS MAINLY ON *bees, wasps and hornets, and their brood. It attacks nests both on the ground and in trees, and scratches out the contents. It removes the stings of adult insects before eating them.*

BREEDING: in spring. Nest of sticks, lined with leafy twigs, high in tree. 1–3 white or creamy-buff eggs, speckled or blotched with chestnut or dark brown, incubated by both parents, especially female, for 30–35 days. Reared by both parents. Fledge at 40–44 days.

OTHER INFORMATION: has distinctive flat-winged flight and long tail.

COMMON KESTREL

Falco tinnunculus

Family: Falconidae

THE COMMON KESTREL IS HIGHLY *adaptable; it is often found in urban areas, where it hunts along roadside verges and over waste land. It hovers as it scans the ground for prey, then swoops down to seize it.*

DISTRIBUTION: breeds on cliffs, buildings and moorland throughout most of Europe and Asia, except the far north, and Africa. Northern populations migrate south in winter.

SIZE: L 31–35 cm (12–14 in); WT 117–299 g (4.1–10.6 oz).

FORM: has long, pointed-wings and long tail. Color of male chestnut with darker spots and black tail bar; head and tail bluish-gray. Female and juvenile lack gray color. Both sexes have dark vertical stripe below the eye, yellow legs, yellowish bill with dark tip and yellow eyes.

DIET: rats, mice and other small mammals; birds, lizards, frogs, insects and earthworms.

BREEDING: in late winter to spring. Lays eggs in tree hollow or rock crevice, on rocky ledge, or in abandoned nest of other birds. 4–9 white or yellowish-buff eggs, heavily speckled with dark reddish-brown, incubated mostly by female for 27–29 days. Chicks brooded and fed by female, with male bringing food. Fledge at 27–39 days.

OTHER INFORMATION: sometimes hunts from a perch, snatching small birds in mid-air.

PEREGRINE FALCON

Falco peregrinus

Peregrine
Family: Falconidae

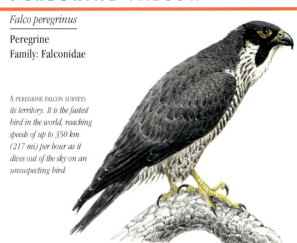

A PEREGRINE FALCON SURVEYS *its territory. It is the fastest bird in the world, reaching speeds of up to 350 km (217 mi) per hour as it dives out of the sky on an unsuspecting bird.*

DISTRIBUTION: open country world-wide, except polar regions, deserts and some islands.

SIZE: L 36–48 cm (14–19 in); WT 611–952 g (1.3–2.1 lb); female much larger than male.

FORM: has long, pointed wings and tapering tail. Color dark grayish-blue above, with whitish to reddish-buff underparts heavily barred with black; legs feathered ; dark mustache, extends around eye; white chin and cheeks; legs, cere and bare skin around eye yellow; bill dark gray. Male darker than female.

DIET: mainly birds.

BREEDING: in spring/summer. Lays eggs in unlined hollow on cliff ledge, high rock or building. 2–6 creamy or buff eggs, heavily marked with chestnut or red, gray and purple, incubated by both parents for 31–38 days. Fledge at 45 days.

OTHER INFORMATION: knocks birds out of sky with talons, then catches them as they fall.

BARN OWL

Tyto alba

Common barn owl
Family: Tytonidae

THE BARN OWL HUNTS AFTER DARK, *using its acute sense of hearing, flying low over the ground or hovering. Its large facial disks of flattened feathers act as sound dishes. The soft plumage muffles the sound of its approach.*

DISTRIBUTION: open country with scattered trees in North and South America, from the United States to the Strait of Magellan, from western Europe to the Black Sea, Africa outside the Sahara, the Middle East, southern Asia and Australia.

SIZE: L 33–35 cm (13–14 in); WT 250–568 g (8.8–20.1 oz).

FORM: a pale owl with a large head, heart-shaped face, dark eyes and long legs. Color reddish gold to rusty-brown above, mottled with gray and buff; underparts, leg feathers and face white; underwings pale; feet brown; bill yellowish to buffish.

DIET: rodents and birds.

BREEDING: in late winter or spring. Nests in tree hollow, rock crevice, abandoned nest or old building. 3–11 white eggs, incubated by female for 32–34 days. Chicks cared for by both parents. Fledge at 60 days.

OTHER INFORMATION: call a prolonged, ghostly screech.

ELF OWL

Micrathene whitneyi

Family: Strigidae

DISTRIBUTION: cactus deserts, dry open woodlands and streamside thickets with trees to 1,740 m (5,700 ft) from the southwest United States to Mexico; some birds from Texas and California overwinter in Mexico.

SIZE: L 13.5–14.5 cm (5–6 in); WT 36–44 g (1.27–1.6 oz).

FORM: a small owl with a short tail. Color mottled brown, gray, black and white, with white stripes on wing and mantle; underparts reddish-brown to whitish, striped; facial disk has white "eyebrows" and dark rim; tail barred; legs, cere and facial skin grayish; bill black, yellowish at base; eyes light orange in male, yellow in female.

DIET: almost exclusively insects; also takes mice and lizards.

THE ELF OWL IS THE SMALLEST OWL IN THE WORLD. *It is often seen looking out of an old wood-pecker hole in a cactus stem. It catches insects by hovering over the ground or vegetation, or by swooping on flying insects from a perch.*

BREEDING: in spring, in desert areas. Nest an unlined cavity in an old wood-pecker nest in a giant cactus or tree. 2–5 white eggs incubated by both parents for 21 days. Chicks cared for by both parents. Fledge at 33–34 days.

OTHER INFORMATION: pair often make a duet of a series of puppy-like barks, whines and yips.

BARKING OWL

Ninox connivens

Winking owl
Family: Strigidae

DISTRIBUTION: forests, woodlands and scrub, especially near open country, across mainland Australia, excluding the central deserts, and some coastal islands; New Guinea and the Moluccas.

SIZE: L 40–45 cm (15.6–17.6 in); WT 462 g (16.3 oz); male much larger than female.

FORM: a medium-sized owl, smoky-brown above, whitish below with bold streaks of dark gray or rust; white spots on wings; legs feathered; patch around eye pale grayish-brown; legs and eyes yellow; bill blackish.

DIET: insects and other small animals.

THE BARKING OWL IS NAMED FOR ITS CALL, A RAPID *dog-like "wuk-wuk", increasing in volume until it carries a long way. In the breeding season it also makes a long, loud, eerie sobbing scream.*

BREEDING: in spring. Nests in tree hollows up to 10 m (33 ft) above ground, lined with rotting wood chippings. 2–3 white eggs, incubated by female for 37 days. Chicks cared for by both parents. Fledge at 35 days.

OTHER INFORMATION: lives in pairs. Roosts by day in tree. Hunts both in trees and on the ground.

CONSERVATION STATUS: not at risk, but not common.

NORTHERN EAGLE OWL

Bubo bubo

Eurasian eagle owl, Eagle owl
Family: Strigidae

DISTRIBUTION: in most habitats, especially rocky areas and forests, across Europe (excluding British Isles), Asia, northern Africa and Middle East.

SIZE: L 60–75 cm (23.5–29.5 in); WT 1.84–4.2 kg (4.1–9.3 lb).

FORM: very large owl; color reddish-brown above, with dark brown markings; wings and tail barred; underparts reddish-buff, with dark streaks and wavy lines; facial disk paler than rest of head; bill black; eyes deep orange. Ear tufts very prominent; legs and feet feathered.

DIET: mammals up to the size of foxes and deer fawns; birds (including other birds of prey); reptiles, amphibians; some insects.

THE NORTHERN EAGLE OWL IS A LARGE, POWERFUL OWL *with a slow, deliberate flight. Its deep, double-hooting call can be heard over one kilometer (0.6 mi) away. Pairs often perform duets.*

BREEDING: in spring. Nest a shallow unlined scrape on ground, cliff ledge, rock crevice, cave, hollow tree, abandoned nest of another bird, even on Egyptian pyramids. 1–6 white eggs, incubated by female for 34–36 days. Young leave nest at 6–10 weeks; fledge at 14 weeks.

OTHER INFORMATION: solitary. Sometimes hunts by day. Roosts in hollow trees, on branches close to tree trunks, or in rock crevices.

SNOWY OWL

Nyctea scandiaca

Family: Strigidae

THE DENSE WHITE PLUMAGE OF THE *Snowy owl, which even covers its feet, camouflages it against the snowy landscape and insulates it against the cold.*

DISTRIBUTION: breeds throughout the Arctic, on tundra and moorlands; winters farther south, as far as temperate zone, around coasts, marshes and sand dunes.

SIZE: L 53–66 cm (21–26 in); WT 1.61 –2.95 kg (3.6–6.5 lb); female larger than male.

FORM: large snow-white owl, with brownish bars and spots near tips of feathers on wings and tail; underparts faintly barred with very pale brown; legs and feet feathered; bill blackish-brown; eyes yellow. Female has much heavier, darker bars and spots.

DIET: mainly lemmings and birds; also hares and other small mammals.

BREEDING: in spring. Female makes a scrape on the ground, among mosses or stones, sometimes lined with moss and feathers. 4–10 white eggs, incubated by female for 32–37 days. Young brooded by female at first; male brings food. Leave nest after 2.5–3.5 weeks. Fledge at 8–9 weeks.

OTHER INFORMATION: hunts by day as well as by night. Call a single hoot; barks and cackles at nest site. Perches on rocks or on good view points to survey its surroundings for prey.

EURASIAN TAWNY OWL

Strix aluco

Tawny owl
Family: Strigidae

WHEN THREATENED, THE TAWNY OWL FLUFFS UP ITS *feathers to make itself look as large and menacing as possible, and hisses at its attacker.*

DISTRIBUTION: open woodland, parks and gardens across Europe, northwest Africa and parts of Asia to southern China up to 2,750 m (9,000 ft).

SIZE: L 37–39 cm (14.5–15 in); WT 385–620 g (13.6–21.9 oz).

FORM: squat, medium-sized brown owl; upperparts reddish-brown, mottled and streaked with dark brown; white patches and dark bars on wings; tail barred; underparts buff, with darker streaks and faint bars; facial disks brown; legs feathered; bill pale greenish-yellow; eyes bluish-black.

DIET: mainly rodents and small birds.

BREEDING: in spring. Nest a shallow unlined depression in a tree hole, rock crevice or abandoned nest of another bird; sometimes on buildings or in nest boxes. 1–7 white eggs, incubated by female for 28–30 days. Chicks fledge at 32–37 days.

OTHER INFORMATION: roosts in tree by day. Call a short and long hoot: "to-whit to-whoo".

WHITE-FACED SCOPS OWL

Otus leucotis

White-faced owl
Family: Strigidae

LARGE, BRIGHT ORANGE EYES SET IN A *white face and emphasized by black "eyeliner" give the White-faced scops owl its fierce look. When threatened, the owl opens them even wider and snaps at its attacker with its bill.*

DISTRIBUTION: in woodlands, thorn scrub and long grasslands of lowland Africa south of Sahara, except forests of Congo basin.

SIZE: L 28 cm (11 in); WT 204 g (7.2 oz).

FORM: smallish owl with ear tufts and large orange eyes set in striking large white facial disks, bordered with black. Upperparts, wings and tail finely mottled brownish-gray, with fine bars of darker gray; underparts less mottled, with more obvious dark streaks; crown and nape black; bill bluish.

DIET: small rodents, small birds, large insects.

BREEDING: season variable. Uses abandoned nests of other birds or unlined hole in tree. 2–4 white eggs, incubated by female for about 25 days. Chicks cared for by both parents even after they leave the nest at 3 weeks; do not fledge until a little later.

OTHER INFORMATION: lives alone or in pairs. Sometimes hunts by day. Roosts in trees near water. Call of male a two-note·hoot "cuc-coo"; female a quavering "wh-h-h-roo".

EURASIAN SWIFT

Apus apus

Common swift, Swift
Family: Apodidae

THE EURASIAN SWIFT SPENDS *almost its whole life in the air, even mating in flight. Its small bill has a wide gape for catching insects on the wing. It skims low over the water surface to drink.*

DISTRIBUTION:
breeds throughout Europe, northwest Africa, and parts of Central Asia and the Far East. Migrates south.

SIZE: L 14 cm (5.5 in); WT 31–43 g (1.1–1.5 oz).

FORM: fast-flying bird with long, curved, tapering wings and short forked tail. Plumage dark grayish-black all over, with whitish throat; legs and bill blackish; eyes dark brown.

DIET: insects.

BREEDING: in late spring and summer. Nests in colonies under eaves or roofs of buildings, in rock crevices or woodpecker holes. Shallow cup-shaped nest made of plant material and feathers, collected from the air, glued together with saliva. 2–4 white eggs, incubated by female for 14–20 days. Chicks fed by both parents. Fledge at 5–9 weeks.

OTHER INFORMATION: outside breeding season, even sleeps on the wing. Seldom alights. Noisy in breeding season, when large groups chase each other and shriek. Young can survive for long periods without food or brooding, becoming torpid.

CRESTED TREE SWIFT

Hemiprocne longipennis

Family: Hemiprocnidae

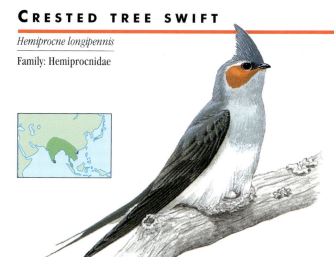

DISTRIBUTION: forest edges, open woodland, parks and gardens from India across southeast Asia to IndoChina. Some populations migrate east or south in winter.

SIZE: L 15 cm (6 in); WT 28–30 g (0.99–1.1 oz).

FORM: a large swift, with deeply forked tail, and crest that is erected when perched. Color dark bluish-gray, with whitish underparts and under-tail coverts; underwing gray; may have fine whitish eyebrow. Male has chestnut cheek patch that extends to throat.

DIET: insects.

A CRESTED TREE SWIFT INCUBATING ITS SINGLE EGG IN *a very flimsy nest. The adult is actually supporting itself on the branch. Unlike true swifts it will sit on branches to rest and preen.*

BREEDING: nest a flimsy basket of leaves and bark glued together with saliva and cemented to side of a branch. Single grayish-white egg fits closely inside nest. Incubated by both parents, which actually rest on branch, so nest does not have to take their weight. Incubation and nestling periods unknown.

OTHER INFORMATION: lives in pairs, which defend a territory; may form loose flocks when hunting. Most active at dusk and dawn.

RIVER KINGFISHER

Alcedo atthis

Common kingfisher, Kingfisher
Family: Alcedinidae

THE RIVER KINGFISHER DEFENDS
*a stretch of water up to 5 km
(3 mi) long. It usually hunts
from a perch overlooking
the water, plunging into the
water, then returning to its
perch to swallow its prey.*

DISTRIBUTION: by
fresh water from
Baltic states, west-
ern Europe and
North Africa to cen-
tral and southeast
Asia and the south-
west Pacific. Northern
birds winter farther south,
often along coasts.

SIZE: L 16 cm (6 in); WT 23–
33 g (0.8–1.2 oz).

FORM: short, stumpy body with short
tail, large head and long, flattened,
pointed bill. Upper parts iridescent
cobalt blue to emerald green; underparts light chestnut; wide chestnut
stripe from bill through eye; throat
white; legs red; bill black, reddish at
base; eyes dark brown.

DIET: mainly fish.

BREEDING: in spring. Pair make tunnel
up to 1 m (3.3 ft) long, in bank near
water, with round nest chamber up to
15 cm (5.9 in) across. No nest material
– becomes lined with fish bones. 4–8
white eggs, incubated by both parents
for 19–21 days. Fledge at 23–27 days.

OTHER INFORMATION: sometimes hovers over water. Call a shrill "cheee" or
"cheekeee". Song a whistling trill.

LESSER PIED KINGFISHER

Ceryle rudis

Pied kingfisher
Family: Alcedinidae

A PIED KINGFISHER HOVERS WHILE IT *scans the water for prey. Since it does not need a perch, it may hunt far from the water's edge.*

DISTRIBUTION: from Asia Minor, the Middle East and Africa south and east of the Sahara to southern Asia.

SIZE: L 28 cm (11 in); WT 68–100 g (2.4–3.5 oz).

FORM: stocky bird with short neck and long, dagger-shaped bill. Striking black-and-white plumage: upperparts black, with feathers edged and streaked white; underparts white with two black breast bands, the upper one broad and incomplete; legs and bill black; eyes dark with white rims. Female lacks lower breast band. Male has crest.

DIET: mainly small fish; also aquatic invertebrates.

BREEDING: in spring. Pair bore a tunnel over 1.8 m (6 ft) long in a bank near water, sloping up to a nest chamber, which gradually becomes lined with fish bones. 3–6 white eggs, incubated by both parents. Young cared for by both parents. Breeds in colonies. Up to 4 non-breeding adults, not necessarily related to breeding pair, may help rear the young.

OTHER INFORMATION: a gregarious bird with a complex social life.

LAUGHING KOOKABURRA

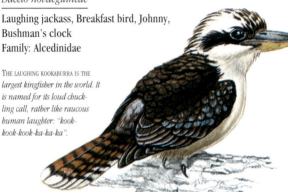

Dacelo novaeguineae

Laughing jackass, Breakfast bird, Johnny,
Bushman's clock
Family: Alcedinidae

THE LAUGHING KOOKABURRA IS THE
*largest kingfisher in the world. It
is named for its loud chuck-
ling call, rather like raucous
human laughter: "kook-
kook-kook-ka-ka-ka".*

DISTRIBUTION: in open forest, cultivated land with trees, parks and gardens of
east and southeast Australia; introduced to southwest Australia and
Tasmania.

SIZE: L 40—45 cm (16—18 in);
WT 305 g (10.8 oz).

FORM: stocky bird, with longish tail,
short neck, large head with stout,
pointed bill. Upperparts blackish-brown
mottled with pale blue on shoulders;
white patch in center of wing; tail red-
dish-cinnamon, barred with black.
Underparts white; head whitish with
blackish-brown streaked patch on
crown and dark brown eyestripe; legs
greenish-gray; bill black above,
creamy-white below; eyes brown.

DIET: fish, small lizards, snakes, small
birds, eggs, crabs, insects, grubs.

BREEDING: September to January.
Nests in unlined holes in trees, termite
mounds or banks, or in buildings. 2–4
white eggs, incubated by both parents
for 25–26 days. Chicks reared by both
parents, often helped by young from
previous year. Fledge at 4–5 weeks.

OTHER INFORMATION: kills large
snakes by dropping them from a height
or battering them on a branch.

EUROPEAN NIGHTJAR

Caprimulgus europaeus

Nightjar, Goat sucker
Family: Caprimulgidae

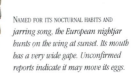

DISTRIBUTION: breeds in open woodlands, heaths and deserts across Europe, southern Scandinavia and Asia as far as Afghanistan and Lake Baikal. Winters in Africa south of the Sahara.

NAMED FOR ITS NOCTURNAL HABITS AND jarring song, the European nightjar hunts on the wing at sunset. Its mouth has a very wide gape. Unconfirmed reports indicate it may move its eggs.

SIZE: L 28 cm (11 in); WT 56–85 g (2–3 oz).

FORM: short-legged bird with small, flattened head, short bill, long tail and wings. Color grayish-brown, spotted and barred with tawny and dark brown. Male has white patches on wingtips and outer tail feathers, obvious in flight.

DIET: insects.

BREEDING: in late spring and summer. Lays eggs in a shallow scrape on the ground, usually among dead leaves and twigs. 2 creamy to grayish-white eggs, spotted, blotched or streaked with yellowish-brown or dark brown; incubated by female by day, male by night, for 18 days. Reared by both parents. Fledge at 16–18 days. Independent at 31–34 days. Female may start second brood while male looks after first one.

OTHER INFORMATION: roosts by day on ground or lying along a branch. Male has wing-clapping display flight, with long, rising-and-falling call.

TAWNY FROGMOUTH

Podargus strigoides

Mopoke, Morepork
Family: Podargidae

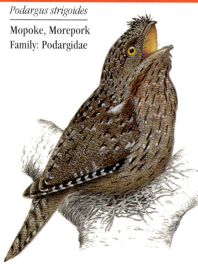

THE TAWNY FROGMOUTH GETS
*its name from its broad,
slightly hooked bill, which
resembles a frog's mouth.
The effect is compounded
by its large yellow eyes.*

DISTRIBUTION: in woodlands and
open country with trees throughout
Australia, including Tasmania and
other coastal islands.

SIZE: L 35–53 cm (14–21 in);
WT 350 g (12.4 oz).

FORM: large bird with long wings and
tail, squat flattened head, very wide bill
and short legs. Color mainly gray, often
tawny or reddish on wings and back,
heavily mottled and marbled with pale
gray or tawny, with dark brown streaks;
tail faintly barred; legs olive-brown; bill
dark brown; eyes bright orange-yellow.

DIET: insects and other small animals.
Hunts from look-out post.

BREEDING: August to February, in tree
fork up to 10 m (33 ft) above ground,
on broken stump or in abandoned nest.
Nest a loose platform of twigs, thinly
lined with leaves. 2–3 white eggs, incubated by female for about 30 days.
Young cared for by both parents. Fledge
at about 35 days.

OTHER INFORMATION: nocturnal. By
day it sits up tall and thin, points bill
toward sky and closes eyes, so that it
looks like an old tree stump.

EUROPEAN BEE-EATER

Merops apiaster

Family: Meropidae

DISTRIBUTION: breeds in open country, cultivated land, scrub and woodland, especially near rivers, from southern Europe and northwest Africa across the Middle East and Asia to northern India, and in South Africa. Winters in Africa.

SIZE: L 25–27 cm (10–10.5 in); WT 56.6 g (2 oz).

FORM: brilliantly colored bird with pointed wings, down-curved bill, square tail with long central feathers, and small feet with partly-joined toes. Color bluish-green with chestnut head and mantle shading to golden yellow on back; wings, tail and underparts bluish-green; throat yellow, with thick black border blending into a distinct black eyestripe; legs and bill blackish-gray; eyes brown.

THE EUROPEAN BEE-EATER PERCHES on posts, wires and trees to look for prey, then glides out rapidly to seize it in mid-air. It batters the insects to death on its perch before swallowing.

DIET: bees and other insects.

BREEDING: in spring and summer, in large colonies. Pair excavate a tunnel, up to 2.7 m (9 ft) long, into a bank, sloping down to a round unlined nest chamber, soon lined with insect remains. 4–10 white eggs, incubated by both parents for 20 days. Chicks cared for by both parents. Fledge at 20–25 days.

OTHER INFORMATION: gregarious. Flight swallow-like. Call a musical "quilp".

EUROPEAN ROLLER

Coracias garrulus

Family: Coraciidae

DISTRIBUTION: breeds in woodland or open areas with trees, from Iberia and southern France, central and eastern Europe to northern India and western Siberia. Winters in tropical Africa.

SIZE: L 30–32 cm (12–12.5 in); WT 127–160 g (4.5–5.6 oz).

FORM: heavily-built bird with stout, slightly hooked bill. Color bluish-green with chestnut back; bright blue wing-patch visible in flight; primaries and tail-tip black; central tail feathers greenish-brown; eyestripe black; legs yellowish; bill brownish; eyes brown.

DIET: mainly insects; also scorpions, frogs, lizards, small birds, snails, fruit.

THE EUROPEAN ROLLER GETS ITS NAME FROM THE *male's dramatic courtship display, in which he tumbles through the sky in twisting dives, to display his colors, at the same time making rattling calls.*

BREEDING: in spring to summer. Nests in tree holes, rock crevices, old buildings or abandoned nests. Nest a hollow, unlined or thinly lined with feathers and debris. 4–7 white eggs, incubated by both parents for 18–19 days. Young hatch helpless. Fledge at 26–28 days, but are fed by parents for some time longer.

OTHER INFORMATION: hunts from a post, branch or telephone wire, swooping down to seize insects in mid-air or on ground.

HOOPOE

Upupa epops

Family: Upupidae

A HOOPOE SPREADS ITS WINGS AND TAIL IN A *defensive display as a bird of prey passes overhead. The hoopoe's name comes from its call, a deep "hoop-hoop-hoop".*

DISTRIBUTION: breeds in woodlands, orchards and parks across Europe (except in north), central and southern Asia and Africa (except Sahara). Northern birds winter farther south, in open country.

SIZE: L 28 cm (11 in); WT 41−83 g (1.4−2.9 oz).

FORM: large crest with barred edge, erected when excited; background color pinkish-brown; wings and tail heavily barred in black-and-white; lower belly streaked with dark brown; rump white; legs slatey-gray; eyes brown. Bill large, slightly down-curved.

DIET: mainly insects, especially crickets.

BREEDING: usually in spring. Nests in tree holes, buildings or nest boxes. 5−12 gray, pale yellow or olive eggs, incubated by female for 16−19 days; male brings food. Chicks reared by both parents. Fledge at 20−27 days.

OTHER INFORMATION: runs after prey. Swift, fluttering flight. Likes dust baths.

NORTHERN FLICKER

Colaptes auratus

Common flicker
Family: Picidae

A NORTHERN FLICKER DISPLAYS TO
*its mate. It is named for its
call, a shrill, repeated "flic",
descending to "kee-oo".*

DISTRIBUTION: grasslands and open
woodlands throughout North America,
from the boreal forest to Nicaragua.
Northern populations winter farther
south.

SIZE: L 25.5–36 cm (10–14 in);
WT 92–167 g (3.2–5.9 oz).

DIET: mainly ants; also fruits, berries.

FORM: color of upperparts brown,
barred with darker brown; underparts
paler, spotted blackish-brown, with
black crescent under chin; rump white,
conspicuous in flight; underwings
bright yellow or reddish, according to
locality; head pattern also varies with
place – some birds have gray nape,
gray cheeks and red or black mustache;
legs gray; bill dark gray; eyes dark
brown; female lacks mustache.

BREEDING: May to August, in hollow
tree, stump or cactus stem up to 30 m
(100 ft), bored by both birds. May use
bird box. 3–14 white eggs, incubated
by female by day, male at night, for
17–18 days. Chicks cared for by both
parents. Fledge at 25–28 days.

OTHER INFORMATION: often feeds on
ground. Flight undulating. Drums with
bill on trees, inside the nest hole and
even on tin roofs.

GREEN WOODPECKER

Picus viridis

Yaffle, Yaffingale
Family: Picidae

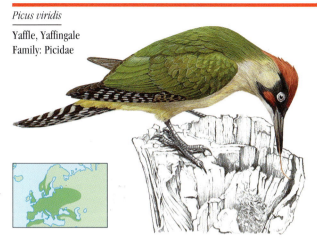

A GREEN WOODPECKER PROBES AN OLD STUMP FOR *insect larvae and spiders. It will also feed on the ground at ant nests, well camouflaged against the green grass.*

DISTRIBUTION: woodlands, forests, parks and gardens from southern Scandinavia, the British Isles (excluding Ireland), western Europe and North Africa to Turkey, Iran and western Russia and the Caspian Sea.

SIZE: L 30 cm (12 in); WT 138–190 g (4.9–6.7 oz).

FORM: color of upper parts bright green, with yellow rump, conspicuous in flight; underparts paler; nape red; crown red in male; eyestripe wide, black; mustache red, edged with black; female's mustache all black; legs and bill dark gray; eyes yellow.

DIET: insect larvae and spiders living under bark; also ants, berries, fruits, seeds; sometimes raids beehives.

BREEDING: in spring. Nest an unlined tree cavity about 38 cm (15 in) deep. 4–9 white eggs, incubated by both parents for 18–19 days. Chicks cared for by both parents. Fledge at 18–21 days.

OTHER INFORMATION: undulating, flap-and-glide flight. Loud, laughing call. Forages on tree trunks, moving in jerky hops, supported by its tail. Solitary outside breeding season.

PILEATED WOODPECKER

Dryocopus pileatus

Family: Picidae

DISTRIBUTION: damp forests with large trees in southern Canada, eastern and northwestern United States.

SIZE: L 38–48 cm (15–19 in); WT 250–309 g (8.8–10.9 oz).

FORM: large black woodpecker, with red cap, white chin, white stripe above and below eye, the latter extending to flanks; white wing lining conspicuous in flight; legs blackish; bill grayish; eyes orange. Male has more extensive cap, always raised.

DIET: ants, termites and other insects; also fruit, nuts, acorns.

BREEDING: in spring. Bores nest hole 25–60 cm (10–24 in) deep in tree

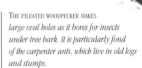

THE PILEATED WOODPECKER MAKES *large oval holes as it bores for insects under tree bark. It is particularly fond of the carpenter ants, which live in old logs and stumps.*

trunk or telegraph pole, up to 20 m (70 ft) above ground. 3–5 white eggs, incubated by female by day, male at night, for about 18 days. Young cared for by both parents. Chicks fledge at 22–26 days.

OTHER INFORMATION: sometimes feeds on ground. Strong, direct flight. Call a loud, repeated rising and falling "wuck-a-wuck-a". Sound of drumming with beak carries a long distance.

NORTHERN WRYNECK

Jynx torquilla

Wryneck, Eurasian wryneck
Family: Picidae

THE WRYNECK IS NAMED FOR ITS *defensive display while at the nest. It will twist its neck into strange positions, hissing like a snake.*

DISTRIBUTION: breeds in deciduous woodlands, parks and gardens throughout much of Europe and northwest Asia, excluding the British Isles and the far north; also in northeast China. Winters farther south, in southernmost Spain, west, central and east Africa south of Sahara, and southeast Asia.

SIZE: L 16–17 cm (6–6.5 in); WT 26–50 g (0.9–1.8 oz).

FORM: color grayish-brown, upper parts mottled, with distinct streak finely marked with darker brown; underparts paler, barred on breast, throat and chin; flecked on belly; tail grayish, barred; grayish band above eye; legs and bill pale yellowish-gray; eyes brown.

DIET: ants and other insects, eggs and nestlings. Has sticky saliva for extracting ants from their nests.

BREEDING: in spring to summer. Nest an unlined hole in a tree, building, bank or nest box. 7–10 white eggs, incubated by both parents, especially female, for 12–14 days. Chicks cared for by both parents. Fledge at 19–21 days. Sometimes has second brood.

OTHER INFORMATION: not very woodpecker-like; perches on branches; hops on ground with tail raised. In courtship display birds face each other shaking heads and displaying pink gapes.

89

OSTRICH

Struthio camelus

Family: Struthionidae

THE OSTRICH IS THE WORLD'S TALLEST AND *heaviest bird – and the fastest. It can run at almost 50 km (30 mi) per hour. Its feet have only two large toes, so there is little friction with the ground as it runs.*

DISTRIBUTION: dry savanna of west, east and south Africa.

SIZE: L 1.75–2.75 m (5.7–9 ft); WT 100–150 kg (220–330 lb); male larger than female.

FORM: large flightless bird, with bare skin on head, neck and thighs. Male black with white wing primaries; tail color varies from whitish to gray or cinnamon-brown; skin of head and neck pink or blue, flushing bright red during courtship; legs and bill horn-colored; eyes brown with large eyelashes. Female pale brown, with pinkish neck and head; wing tips sometimes whitish.

DIET: plant material.

BREEDING: season varies with latitude. Male makes unlined scrape on ground. Several females lay up to 8 yellowish-white eggs in common scrape, incubated by male at night, and by dominant female by day, for 42 days. Nest may contain up to 78 eggs at first. Dominant female often pushes out some of the other females' eggs. Chicks leave nest soon after hatching. Cared for by both parents for several months; often join creches.

OTHER INFORMATION: Call loud and deep, rather like lion's roar.

LESSER RHEA

Pterocnemia pennata

Darwin's rhea
Family: Rheidae

THE LESSER RHEA CANNOT FLY, BUT IT IS VERY FLEET *of foot, with a stride of up to 1.5 m (5 ft). When in danger, it flees, zigzagging as it runs. It may even turn round and leap over the heads of its pursuers.*

DISTRIBUTION: scrublands and grassy steppes of South America, in Patagonia and Andes up to 4,000 m (1,220 ft).

SIZE: L 90 cm (36 in); WT 15–25 kg (33–55 lb); female slightly smaller than male.

FORM: large flightless bird with long legs and neck. Brown with white flecks.

DIET: mainly grasses, herbs and seeds; some insects and small vertebrates.

BREEDING: from September to December. Male fights rivals to acquire a harem of females. He makes a scrape on the ground, about 1 m (3.3 ft) across, thinly lined with plant material. Several hens each lay up to 20 yellowish-green eggs in it, until it contains up to 80 eggs. Incubated by male for about 40 days. Leaves nest only for about an hour around midday to feed and drink. Young well developed on hatching. Cared for by male for several months.

OTHER INFORMATION: forms loose flocks of 50–100 birds outside breeding season. Males become highly territorial in breeding season. When running, it raises one wing and lowers the other, which enables it to make quick changes of direction.

ONE-WATTLED CASSOWARY

Casuarius unappendiculatus

Family: Casuariidae

DISTRIBUTION: lowland forests of New Guinea.

SIZE: L 100 cm (3.3 ft); wt not known, probably about 20.7 kg (46 lb).

FORM: heavy flightless bird with small wings, powerful legs and feet with only three large toes. Color dark brown to black, with bare blue skin on head and 2 folds at base of beak, abruptly changing to red, sometimes yellow, skin on lower neck, with a further fold, and a red or yellow stripe, running toward shoulder; flattened horny "casque" or "helmet" on top of head; legs grayish-black; bill horn-colored; eyes yellowish.

DIET: fruit, seeds and other plant material; small animals.

TWO COLOR FORMS OF THE ONE-WATTLED CASSOWARY are shown here. This species has only one red, sometimes yellow, skin flap on its neck. Cassowaries use their hard, horny "helmets" to push through dense undergrowth on the forest floor.

BREEDING: nest a shallow scrape on forest floor, lined with grass and leaves. 3–8 eggs, pale to dark green, laid by a single female, incubated by male for 49–56 days. Chicks leave nest soon after hatching. Reared by male only.

OTHER INFORMATION: lives alone. If threatened, it will leap up and lash out with the sharp claw of the inner toe, which is up to 10 cm (4 in) long, trying to disembowel its opponent.

EMU

Dromaius novaehollandiae

Family: Dromaiidae

A MALE EMU STANDS GUARD OVER HIS EGGS AND *newly-hatched young. He will probably have to rear them single-handed. While incubating the eggs, he has been fasting. Soon he will be free to feed again.*

DISTRIBUTION: throughout mainland Australia, except in driest deserts, dense forests and urban areas. Nomadic.

SIZE: L 180 cm (71 in); WT 31 kg (68 lb).

FORM: large flightless bird with long, double feathers. Color ranges from grayish-brown to almost black, with whitish ruff at base of neck; skin of face and neck blue; legs dark grayish-brown; bill blackish; eyes yellow, grayish-brown or reddish.

DIET: fruits, seeds (including grains).

BREEDING: from April to November. Nest a platform or circle of leaves, grass, bark or sticks on ground or

under bush or tree. 5–11 eggs, dark grayish-green, incubated by male alone. Chicks leave nest soon after hatching. Cared for by male for about 18 months. Female may stay nearby or leave to mate with another male.

OTHER INFORMATION: lives alone, in pairs, small family groups or large flocks. Male makes deep growling grunts; female makes rattling sounds.

CONSERVATION STATUS: not at risk. Has been persecuted for taking grain.

BROWN KIWI

Apteryx australis

Common kiwi
Family: Apterygidae

DISTRIBUTION: scattered populations in forests of New Zealand

SIZE: L 70 cm (27.5 in): WT 1.72–3.85 kg (3.8–8.5 lb); female larger than male.

FORM: flightless bird with thin feathers rather like mammalian hair, and no visible wings or tail. Bill long and flexible, with nostrils near tip and large, sensitive whiskers around base. Color of plumage and legs brownish; bill yellowish to horn colored; eyes very small.

DIET: insects (especially larvae), spiders, earthworms, berries.

BREEDING: in late winter. Nest a hollow, unlined or lined with leaves and

THE BROWN KIWI IS UNUSUAL AMONG BIRDS IN *having an acute sense of smell and nostrils near the tip of its bill. It probes the ground for worms and insects. Kiwis are named for the shrill "kee wee" call of the male.*

leaf mold, in a burrow, under tree roots or in a hollow log. 1–2 white eggs, incubated by male for 65–85 days. Chicks hatch well developed. Leave nest after 5 days, living on egg yolk. Cared for by male for some time.

OTHER INFORMATION: nocturnal. Call of female a hoarse "kurr kurr".

ELEGANT CRESTED TINAMOU

Eudromia elegans

Elegant tinamou, Crested tinamou, Martineta tinamou
Family: Tinamidae

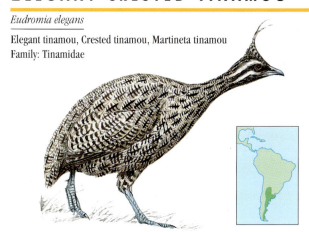

DISTRIBUTION: scrub, woodlands, dry steppes and semi-arid grasslands of South America, from Bolivia, Argentina and Paraguay to southern Chile, up to 3,500 m (11,500 ft).

SIZE: L 40 cm (16 in): WT 660 g (23.3 oz).

FORM: plump bird with short tail and distinctive slender, forward-curving black crest. Only 3 toes on each foot, perhaps to reduce friction with ground for running across open country. Color brownish, marbled and finely streaked and barred with dark brown, bars wider on wings and tail; head finely striped dark brown and cream, crest black; legs, bill grayish; eyes brownish.

THE ELEGANT CRESTED TINAMOU LIVES ON THE FOREST *floor, seldom taking flight. If unable to escape danger on foot or by hiding, it may suddenly fly up noisily, rather like a pheasant.*

DIET: mainly small fruits, seeds; also buds, shoots, flowers, roots, insects, worms, mollusks.

BREEDING: nest a scrape on the ground, made by male. Male attracts 2 or more females by calling. They lay 1–16 green eggs in the scrape, incubated by male for 16–20 days. Chicks hatch well-developed; leave nest soon after hatching. Cared for by male.

OTHER INFORMATION: lives in flocks of up to 100 birds. Likes to dust-bath. Has flute-like whistling call.

ORANGE-FOOTED SCRUB FOWL

Megapodius reinwardt

Australian scrub fowl, Scrub fowl,
Scrub hen, Jungle-fowl
Family: Megapodiidae

THE ORANGE-FOOTED SCRUB FOWL IS
*renowned for the huge nest
mounds it builds, up to 5 m
(16.4 ft) high and 12 m (39 ft)
across. It acts as an incubator
for the eggs, as the rotting
vegetation gives off heat.*

DISTRIBUTION: dense thickets near water along rivers, coasts and island shores of northern Australia, New Guinea, Indonesia, Philippines, Solomon Islands and New Hebrides.

SIZE: L 45.5 cm (17.7 in); WT 0.6–1.02 kg (1.3–2.25 lb).

FORM: chicken-like bird with large feet. Color various shades of brown, darker on crown and nape, reddish on tail coverts, blackish on tail, with bluish-gray underparts and back of neck; short crest on head; legs bright orange; bill reddish-brown; eyes brown, ringed by bare yellow skin.

DIET: seeds, fruits, berries, roots, insects and other invertebrates.

BREEDING: from August to March. Makes a huge mound of earth or sand and vegetation, used and added to for several years. 3–13 yellowish-white to pale pink eggs laid singly in deep holes in the mound and covered over. Chicks hatch well-developed and are independent immediately.

OTHER INFORMATION: lives in pairs. Voice a loud, harsh double crow; noisy at dawn and dusk, and at night. When alarmed, runs away or flies to a low branch and "freezes".

WESTERN CAPERCAILLIE

Tetrao urogallus

Capercaillie
Family: Tetraonidae

A MALE CAPERCAILLIE DISPLAYS *to females at a communal display ground, fanning his tail, uttering a rattling call that ends with a sound like the popping of a cork.*

DISTRIBUTION: conifer woods and scrub of northwest Spain, northern and central Europe and the Balkans, east across northern Asia.

SIZE: L 60–87 cm (23.5–34 in); WT 1.7–5.1 kg (3.7–11.1 lb); male considerably larger than female.

FORM: turkey-like bird with feathered legs. Male dark gray, with white on flanks; wings and shoulders chestnut; breast glossy green mottled with white; belly mottled blackish and white; throat and face black; legs brown to gray; bill horn-colored; eyes dark, with red eyebrow. Female brown to buff, mottled with white and black; breast chestnut; eyebrow red.

DIET: buds and shoots of conifers; seeds, grasses, berries, fruits; insects.

BREEDING: in spring. Female makes scrape in undergrowth or at foot of tree, lined with plant material and debris. 5–8 pale yellowish-buff eggs, incubated by female alone for 26–29 days.

OTHER INFORMATION: roosts in trees.

SAGE GROUSE

Centrocercus urophasianus

Family: Tetraonidae

A MALE SAGE GROUSE DISPLAYS
*to the females on a commu-
nal display ground called a
lek. As he fluffs out his chest
feathers and fans his tail, he
utters loud, bubbling popping sounds, aided by the
two yellow throat sacs.*

DISTRIBUTION: sagebrush plains, hills and deserts of North America, from Washington and Saskatchewan to eastern California and western Colorado.

SIZE: L 72 cm (28 in); WT 1.75–3.19 kg (3.9–7 lb); male larger than female.

FORM: sturdy, plump bird with long, pointed tail. Female yellowish-brown, with black belly; legs feathered, grayish; bill dark; eyes dark brown. Male similar, but with black face and throat with white half-collar, and white breast with two yellow air sacs; has yellow wattle over each eye.

DIET: leaves and shoots of sagebrush, legumes in winter; also seeds, berries and insects.

BREEDING: in spring. Males promiscuous. Nest a scrape on the ground, made by female, lined with plant material and debris, often under sagebrush. 7–8 pale olive to buffish-olive eggs, incubated by female alone for 25–27 days. Young hatch well-developed and leave nest almost immediately. Cared for by female. Fledge in 2 weeks.

OTHER INFORMATION: up to 400 males may display at a single lek.

ROCK PTARMIGAN

Lagopus mutus

Ptarmigan
Family: Tetraonidae

A ROCK PTARMIGAN IN WINTER *plumage is well-camouflaged against the snow. In spring, the male changes to summer camouflage later than the female. It has been suggested that this helps him distract predators away from the nest.*

DISTRIBUTION: rocky mountains, usually above 900 m (3,000 ft), across northern parts of North America, Pyrenees, Alps, Europe and Asia.

SIZE: L 35–36 cm (14 in); WT 359–482 g (12.7–17 oz).

FORM: short-legged ground-dwelling bird with blunt tail, feathered legs and feet. Winter plumage all white, with black tail; red wattles over eyes; male has black stripe through eye. Summer plumage of female buffish-brown, finely barred with black; cheeks paler; male in summer has brownish upper parts, breast and flanks mottled with reddish-brown, white throat and belly.

between rocks or bushes, thinly lined with grass, leaves and feathers. 3–12 whitish to creamy-yellow eggs, blotched and mottled with dark brown, incubated by female for 24–26 days; male stays nearby. Young fledge at 10 days.

DIET: leaves, seeds, berries; some insects.

BREEDING: in late spring to summer. Female makes scrape on ground

OTHER INFORMATION: burrows in snow for shelter and food in winter. Call a harsh croak. If eggs or young threatened, will feign injury.

GREATER PRAIRIE CHICKEN

Tympanuchus cupido

Family: Tetraonidae

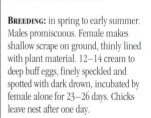

A MALE GREATER PRAIRIE *chicken displays to watching females at a "booming ground" on the prairies, running to and fro and leaping into the air, making loud booming calls.*

DISTRIBUTION: scattered populations on tall-grass prairies, scrublands and cultivated land from south-central Canada to the Gulf of Mexico.

SIZE: L 43–48 cm (17–19 in); WT 0.71–1.14 kg (1.6–2.5 lb).

FORM: large brown grouse with short rounded tail and feathered legs. Color brown, heavily barred all over with dark brown, cinnamon and creamy-white; tail blackish in male, barred in female; feet grayish; bill and eyes dark. Male has yellowish-orange neck sacs and eye wattles, and elongated feathers on nape of neck that can be erected.

DIET: leaves, buds, fruits, seeds (especially grains in winter); also insects.

BREEDING: in spring to early summer. Males promiscuous. Female makes shallow scrape on ground, thinly lined with plant material. 12–14 cream to deep buff eggs, finely speckled and spotted with dark drown, incubated by female alone for 23–26 days. Chicks leave nest after one day.

OTHER INFORMATION: burrows into snow for shelter in winter.

CONSERVATION STATUS: not at risk, but declining rapidly in some areas.

COMMON PHEASANT

Phasianus colchicus

Ring-necked pheasant
Family: Phasianidae

THE MALE COMMON PHEASANT IS A HAND-
some bird, but not well-camouflaged. The spurs on his feet are
used in fights with rival males
during the breeding season.

DISTRIBUTION:
originally woodlands,
marshes and cultivated
areas of central and eastern
Asia; introduced to Europe, North
America, Tasmania, New Zealand.

SIZE: L 52–90 cm (20.5–35.5 in);
WT 0.95–1.32 kg (2.1–2.9 lb); male
considerably larger than female.

FORM: color of male ranges from
bright reddish-brown to dark purplish-
brown, with rufous lower back and
rump and long, pointed, copper-brown
tail; head and neck iridescent dark
green, often with white collar; small ear
tufts; large patch of red skin around
eye. Female mottled rufous and blackish, with shorter tail.

DIET: seeds (including grains), nuts,
acorns, berries, leaves, fruits; insects.

BREEDING: in spring or early summer.
Female makes shallow scrape, unlined
or thinly lined with plant material.
7–15 olive, olive-brown, brown or
bluish-gray eggs, incubated by female
alone for 23–27 days.

LADY AMHERST'S PHEASANT

Chrysolophus amberstiae

Family: Phasianidae

DISTRIBUTION: dense woodland, thorn thickets and bamboo scrub in mountainous regions of southern China, northeast Burma and nearby areas; introduced to England.

SIZE: L 58–68 cm (23–27 in); WT 624–850 g (1.4–1.9 lb).

FORM: male iridescent green with white crown, nape and ruff; feathers edged in black; crest pinkish-red; eye-patch blue; golden rump, fringed by long pinkish-red feathers; tail long, white with black bars and markings; belly white; legs and bill grayish; eyes yellow. Female brownish; underparts buffish-brown with dark bars and mottlings; throat plain cream; eye patch pale blue; eyes brown .

DIET: mainly bamboo shoots; insects, spiders and other small invertebrates.

BREEDING: in spring to early summer; 5–12 creamy or yellowish eggs, incubated by female alone for 22–23 days. Chicks hatch well-developed. Cared for by female.

OTHER INFORMATION: forms flocks of up to 30 birds in winter.

NORTHERN BOBWHITE

Colinus virginianus

Bobwhite quail
Family: Phasianidae

NAMED FOR ITS LOUD "BOB-
*white" call, the Northern
bobwhite quail is seldom
noticed outside the breeding season. Several quail
roost together on the
ground in a tight circle.*

DISTRIBUTION: pine woods, scrublands, cultivated land and urban areas of North and Central America, from the Great Lakes south to Guatemala, and on Cuba.

SIZE: L 20–28 cm (8–11 in); WT 178 g (24.1 oz).

DIET: seeds, berries, grass, leaves, insects.

FORM: a small chunky bird. Color reddish-brown mottled with dark brown; feathers pale-edged; primaries and tail feathers dark brown; flanks reddish-brown streaked with white and black; belly white, barred with black; legs pinkish; bill yellowish; eyes dark. Male has white stripe through eye, and white throat; throat and eyebrow buff instead of white in female.

BREEDING: in spring. Pair make a shallow scrape, lined with plant material and covered in vegetation. 12–16, up to 28, dull white to creamy-white eggs, incubated by female for 18–20 days. Chicks cared for by both parents. Fledge at 2 weeks.

OTHER INFORMATION: form flocks of up to 30 birds in fall and winter.

COMMON TURKEY

Meleagris gallopavo

Wild turkey
Family: Meleagrididae

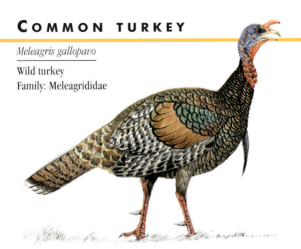

DISTRIBUTION: scattered in open woodlands, forest clearings and scrublands of North America, from Oklahoma and Pennsylvania to southern Mexico.

SIZE: L 86–122 cm (34–48 in); WT 4.22–7.40 kg (9.3–16.3).

FORM: slightly smaller and slimmer than domestic turkey. Color of male iridescent dark bronze-brown, with black-edged feathers; primaries barred with white; black tuft of feathers on breast; skin of head blue and pink, with red wattles; legs light brown, with spurs; bill dark yellowish-brown; eyes dark. Female duller, often without breast tuft.

DIET: seeds (including grains), nuts, fruits; insects and other invertebrates.

THE COMMON TURKEY MALE HAS SPURS ON HIS FEET *for fighting rivals, sometimes to the death. In spring, he will put on a magnificent courtship display, fanning his large tail.*

BREEDING: in spring. Males promiscuous. Female makes shallow scrape, thinly lined with plant material. 8–12, up to 20, pale buff or yellowish eggs, heavily speckled and spotted with brown or purplish-brown, incubated by female for 28 days. Chicks cared for by female. Fledge at 2 weeks.

OTHER INFORMATION: Roosts in trees.

HELMETED GUINEAFOWL

Numida meleagris

Tufted guineafowl,
Hooded guineafowl
Family: Numididae

A HELMETED GUINEAFOWL LAUNCHES *into a defensive display to protect its young. Its fleshy, colorful wattles may act as signals to help keep the flock together.*

DISTRIBUTION: dry scrub, thorn thickets, open woodland, grassland and cultivated land throughout much of Africa south of the Sahara, and in southwest Morocco.

SIZE: L 53–58 cm (21–23 in); WT 1.3 kg (2.9 lb).

FORM: plump gamebird with very small head and neck, and large, horny casque ("helmet") on head. Color blackish-gray, with rows of white spots and streaks; head and neck bare, cobalt blue to bluish-white; cere (fleshy pad at base of bill) and wattles red and blue; tuft of white bristles over nostrils; legs and eyes dark brown.

DIET: insects, earthworms, mollusks, seeds (including grains), bulbs, roots.

BREEDING: season varies with location. Nest a scrape on ground, lined with grass and feathers, in long grass or under vegetation; sometimes shared by 2 females. 6–12, up to 20, cream or yellowish-buff eggs, speckled with brown and white. Incubated by female. Chicks fledge in a few weeks.

OTHER INFORMATION: live in pairs in breeding season, otherwise in flocks.

KORI BUSTARD

Choriotis kori

Giant bustard
Family: Otididae

A CARMINE BEE-EATER HAS HITCHED A LIFT ON *this Kori bustard. It will dart out to catch the insects disturbed by the bustard's stately progress through the grass.*

DISTRIBUTION: thorn bush and grass-lands of eastern and southern Africa.

SIZE: L 100–130 cm (39–51 in); WT 5.9–14.5 kg (13–32 lb).

FORM: very large ground-dwelling bird with long crest on head. Upperparts brownish-buff with fine black mark-ings; flight feathers black, barred and mottled with white; upper wing coverts white with black spots, showing as bar along flank when wings closed; tail black-and-white; underparts white, with collar of small black crescents; legs and eyes yellow; bill horn-colored.

DIET: insects, especially locusts; small vertebrates.

BREEDING: season variable, related to rains. Often breeds communally. No nest. 2 pale greenish-brown eggs, streaked with dark brown, laid on ground and incubated by female for about 4 weeks. Young fed by female at first. Fledge at about 5 weeks.

OTHER INFORMATION: live alone or in pairs. Male puts on spectacular courtship display, raising his tail and swelling his throat, fluffing out his feathers to appear white almost all over, rather like a powder puff.

RED-LEGGED SERIEMA

Cariama cristata

Crested seriema
Family: Cariamidae

A CRESTED SERIEMA WITH A LIZARD IT HAS JUST *caught. Seriemas often catch snakes and lizards, including venomous species. They live mainly on the ground, and if disturbed, will run away rather than fly.*

DISTRIBUTION: dry savanna and other grasslands of South America, from eastern Bolivia and central Brazil to northern Argentina and Paraguay.

SIZE: L 100 cm (39 in); WT 1.4 kg (3.1 lb).

FORM: tall ground-dwelling bird with long legs, long neck and hooked bill, with tuft of bristly feathers at base. Upperparts grayish-brown; wings dark brown with white band; tip of tail barred black-and-white; underparts and cheeks pale gray with fine dark gray mottling; skin of face bluish; bill red; eyes light brown.

DIET: small vertebrates, especially amphibians, lizards and snakes.

BREEDING: builds large nest of sticks in bush or low tree, up to 3 m (10 ft) above ground. 2–3 pale pink eggs, spotted and lined with brown, gray and purple, incubated for 25–26 days.

OTHER INFORMATION: voice a high-pitched yelp. Easily tamed; used by local farmers to warn of intruders.

NAMAQUA SANDGROUSE

Pterocles namaqua

Family: Pteroclididae

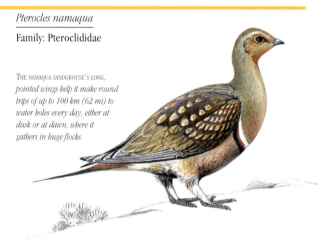

THE NAMAQUA SANDGROUSE'S LONG, *pointed wings help it make round trips of up to 100 km (62 mi) to water holes every day, either at dusk or at dawn, where it gathers in huge flocks.*

DISTRIBUTION: arid regions of southwest Africa, within flying distance of watering holes.

SIZE: L 28 cm (11 in); WT 143–193 g (5–6.8 oz).

FORM: plump ground-dwelling bird with small head, short bill, short legs and long, tapering tail. Male tawny brown to buffish-brown, yellowish around eyes; white and brown crescent on breast; wings mottled grayish-brown and yellow; primaries tipped with black; tail dark; legs grayish, partly feathered; bill grayish; eyes dark. Female more yellowish, and except for cheeks and undertail, streaked and mottled with brown.

DIET: small seeds.

BREEDING: female makes a shallow scrape on ground, lined with plant material and small stones. 3 dull cream, pink or green eggs, spotted and blotched, incubated by both parents for 21–31 days. Young hatch well developed, leave nest almost immediately.

OTHER INFORMATION: the male carries water to the chicks in special belly feathers that can hold a lot of water.

GREATER ROADRUNNER

Geococcyx californianus

Family: Cuculidae

THE GREATER ROADRUNNER PREYS ON REPTILES, *which it pursues on foot at great speed. It will even catch rattlesnakes. A lively, inquisitive bird, its presence is often welcomed on farms.*

DISTRIBUTION: deserts, semi-deserts and scrublands of the southeastern United States and northern Mexico.

SIZE: L 51–61 cm (20–24 in); WT 221–538 g (7.8–19 oz).

FORM: ground-dwelling bird with long, hooked bill; legs long; wings short and rounded; tail very long. Color dark brownish, with feathers edged in pale brown to buff, giving mottled appearance; dark brown crest on head; tail tipped with white; underparts and neck white with dark brown streaks; legs gray; bill yellowish; eyes pale green, rimmed with red.

DIET: lizards, small snakes, insects.

BREEDING: in April and May. Makes nest of twigs in bush, cactus or low tree, lined with grass, leaves, roots, feathers and debris. 3–6 white or cream eggs, incubated by female for 18–19 days. Chicks hatch helpless; cared for by both parents. Fledge at about 16 days. May have 2 broods a year.

OTHER INFORMATION: lives in pairs; defends territory all year. Call a dove-like cooing, descending in pitch.

DUSKY BROADBILL

Corydon sumatranus

Family: Eurylaimidae

DISTRIBUTION: high in the canopy of forests of southeast Asia, Sumatra and northwest Borneo, especially along streams.

SIZE: L 27.5 cm (11 in).

FORM: stout bird with large head, very broad, flattened, hooked bill and short tail; third and fourth toes partly fused. Color dull black, with buffish throat and narrow white band near tip of tail; concealed yellowish-orange streak on back; wing bar broad, white; legs blackish; bill pinkish.

DIET: mainly insects; also spiders, snails and other invertebrates; tree frogs; fruits, seeds, buds.

BREEDING: several birds may help to build a pear-shaped nest of stems, roots, grass and leaves, with an opening

THE DUSKY BROADBILL HAS A PARTICULARLY WIDE BILL, which may help this rather slow-moving bird to capture insects in flight. It also lies in wait until crickets and grasshoppers come within reach, then leaps up to seize them.

at the side, sometimes decorated with spider webs, hung by a plant fiber cord from a branch, usually hanging over a river. 3 speckled eggs, incubated by both parents. Incubation and fledging periods unknown.

OTHER INFORMATION: call a "ki-ip, ki-ip". Lives in noisy parties comprising up to 10 birds.

SCALE-THROATED EARTHCREEPER

Upucerthia dumetaria

Family: Furnariidae

DISTRIBUTION: dry scrublands and hill slopes of southern South America, in southern Peru, western Bolivia, Chile and Argentina.

SIZE: L 25 cm (10 in); WT 30.1 g (1.1 oz).

FORM: large ground-dwelling bird with long tail and long, down-curved bill. Color of upperparts brown, with chestnut tinge to primaries and outer tail feathers, dark brown streaked cheeks and white eyebrows; underparts whitish; feathers of flanks and belly brownish-tipped to give scaly effect.

DIET: insects and other invertebrates.

THE SCALE-THROATED EARTHCREEPER IS A RATHER *inconspicuous ground-dwelling bird that seldom takes to the air. If frightened, it will run away rather than fly.*

The long, down-curved bill is probably used to probe around trunks and logs and investigate cracks and crevices.

BREEDING: pair excavates burrow up to 1.5 m (5 ft) deep in hill slope or earth bank, lined with grasses and roots. 2–3 white eggs, incubated by both parents for about 15 days. Young cared for by both parents. Fledge at about 12 days.

OTHER INFORMATION: forages on the ground.

WHITE-PLUMED ANTBIRD

Pithys albifrons

White-fronted antbird,
White-faced antbird
Family: Formicariidae

THE WHITE-PLUMED ANTBIRD LIVES IN LOW *vegetation near the forest floor. It follows columns of army ants, snatching up insects such as cockroaches and katydids as they flee from the advancing ants.*

DISTRIBUTION: tropical forest undergrowth of South America, from eastern Colombia south to eastern Peru and Brazil north of River Amazon.

SIZE: L 11 cm (4 in); WT 20.8 g (0.7 oz).

FORM: small bird with moderately long bill, hooked at the tip, with notches on the mandibles to help grip prey. Color upperparts gray; head dark gray, with white stripe behind eye; underparts, rump, tail and collar at nape of neck chestnut; legs yellowish-red; bill dark gray; eyes golden-brown. Male has unusual plumes of white feathers on head and chin.

DIET: insects and spiders.

BREEDING: pair for life. Pair builds frail, deep nest of loosely woven plant fibers slung in fork of twig close to ground, lined with leaves and other plant material. 2 white, dark-spotted eggs, incubated by both parents for about 14 days. Young cared for by both parents. Fledge at about 14 days, but remain for much longer with their parents.

OTHER INFORMATION: forages in parties of up to 25 birds.

AMAZONIAN UMBRELLABIRD

Cephalopterus ornatus

Ornate umbrellabird,
Umbrellabird
Family: Cotingidae

A MALE AMAZONIAN UMBRELLABIRD ERECTS HIS *umbrella-like crest and inflates his throat wattle, bobbing forward and down, and making deep booming sounds as he tries to attract a mate. Males usually have a distinct display tree.*

DISTRIBUTION: lowland tropical rainforest of South America, in clearings, along river banks or on islands, in the Amazon basin, southern Venezuela and Guyana.

SIZE: L 51 cm (20 in); WT 380 g (13.4 oz).

FORM: large bird with longish, thick bill, fluffy flat-topped crest and long feathered wattle hanging from throat, largest in male. Color glossy bluish-black; feathers of crest have white shafts; legs slatey-gray; upper mandible gray, lower mandible yellowish-brown; eyes white or pale gray. Female duller than male.

DIET: mainly fruits; some insects.

BREEDING: makes cup-shaped nest, in tree. 1–3 probably buff or olive eggs, spotted and blotched with dark brown and gray, incubated for 19–21 days. Young fledge at 13–15 days.

OTHER INFORMATION: lives mainly in tree tops. Has deeply undulating flight, usually high above ground.

BLUE-WINGED PITTA

Pitta brachyura

Indian pitta, Fairy pitta
Family: Pittidae

DISTRIBUTION: dense forest, scrub and swamps of southeast Asia, from India to Indonesia.

SIZE: L 19–20 cm (7.4–7.8 in); WT 47–66 g (1.8–2.3 oz).

FORM: small bird with plump breast and short tail. Color of upperparts brilliant green; rump and wings mainly cobalt blue, with black, white and green wing-bars; tail barred blue and black at tip; crown blackish-brown, with lighter brown streaks above eyes; throat white; underparts buff, shading to rufous or crimson on flanks; legs yellowish-gray to yellowish-pink; bill dark yellowish-brown; eyes dark brown.

DIET: mainly insects and spiders.

THE BLUE-WINGED PITTA IS A COMMON, BRILLIANTLY *colored little bird that continually bobs up and down and flicks its tail.*

BREEDING: season variable. Builds large, untidy spherical nest of twigs, roots, grasses, leaves and mosses, with side entrance, usually between tree roots near water. 2–5 white or cream eggs, heavily marked with red and purple, incubated by both parents, probably for 15–17 days. Chicks probably fledge at 14–21 days.

OTHER INFORMATION: flight whirring.

ROYAL FLYCATCHER

Onychorhynchus coronatus

Family: Tyrannidae

A MALE ROYAL FLYCATCHER DISPLAYS HIS SPECTACULAR *crest and fans his tail. Normally the crest is scarcely noticeable. The broad bill is fringed with long bristles, which probably help it trap insects in flight.*

DISTRIBUTION: edges of moist forest, near rivers and streams, of northern South America to Bolivia and south-eastern Brazil.

SIZE: L 16.5 cm (6.5 in); WT 9.7–16.2 g (0.34–0.57 oz).

FORM: small, slender bird with broad, flat, slightly hooked bill and wedge-shaped head (due to folded crest). Color tawny-brown; wings darker, with white spots on coverts; tail rufous; head has large crimson crest, smaller and orange in female, tipped with black and blue disks; throat whitish; under-parts pale buff, faintly barred with black on breast and flanks; legs and bill brown; eyes golden-brown.

DIET: insects, caught on the wing by darting out from perch.

BREEDING: pairs for life. Makes slender pouch-like nest up to 2 m (6.6 ft) long, suspended from a branch or vine, often over water. 2 whitish eggs, incubated by female for probably 14–23 days; male stands guard. Young probably fledge at about 14–23 days, but will remain with both parents for a long time afterward.

OTHER INFORMATION: call a loud, mellow whistle.

GREAT CRESTED FLYCATCHER

Myiarchus crinitus

Family: Tyrannidae

DISTRIBUTION: breeds in open woodland and forest edges of eastern North America, from southern Canada to Texas and the Gulf coast; winters farther south, from Florida to South America.

SIZE: L 18–20 cm (7–8 in); WT 27.2–39.6 g (1–1.4 oz).

FORM: small bird with longish tail, crest on head and broad, slightly hooked bill with bristles at base. Color of upperparts brownish; throat and breast grayer; tail feathers dusky, with reddish inner webs; lower breast and flanks brownish; belly and undertail coverts bright lemon yellow; legs grayish-black; bill yellowish, darker at tip; eyes dark.

THE GREAT CRESTED *flycatcher darts out from a favorite perch to catch insects on the wing. It defends its territory first by loud, whistling calls, and if necessary by chasing off intruders.*

DIET: insects.

BREEDING: in spring. Nests in treehole, hollow post or nest-box up to 21 m (70 ft) above ground. Nest a platform of dead leaves, twigs, feathers, hair and debris. 4–5 creamy-buff or ivory-yellow eggs, heavily streaked and scrawled with purple, incubated by female for 13–15 days. Young fledge at 14–15 days, but stay with parents.

OTHER INFORMATION: has curious habit of lining nest with snake skins.

LONG-TAILED MANAKIN

Chiroxiphia linearis

Family: Pipridae

DISTRIBUTION: dry forest edges, woodlands and mangroves of Central America, from southern Mexico to Costa Rica, mainly in lower levels.

SIZE: L 11–22 cm (4–8.5 in); WT 16.8–19.1 g (0.59–0.67 oz); male twice as large as female.

FORM: small bird with short, slightly hooked bill; central tail feathers wire-like up to 12 cm (5 in) long. Color of male black, with bright pale blue back and red crown; very long central tail feathers; legs orange; bill black; eyes dark brown. Female dull green, with slightly paler underparts and shorter tail feathers.

THE MALE LONG-TAILED MANAKIN HAS VERY LONG TAIL *feathers. Males perform group courtship displays, cartwheeling over each other and "mewing" as they try to attract the females.*

DIET: fruits and insects.

BREEDING: starts April or May. Makes suspended cup nest of leaves and moss in a bush up to 3 m (10 ft) above ground. 2 cream eggs, heavily streaked with brown and gray, especially at larger end, incubated by female for 18–19 days. Chicks cared for by female alone. Fledge at about 14 days.

OTHER INFORMATION: flies rapidly. Snatches both fruits and insects during short flights from a perch.

RIFLEMAN

Acanthisitta chloris

Family: Xenicidae

Distribution: forests and scrublands of New Zealand and nearby islands, except northern part of North Island; also cultivated areas.

Size: L 8 cm (3 in); WT 5.0–8.5 g (0.18–0.3 oz).

Form: small, stout bird that flicks wings while perched. Color of male: upperparts bright yellowish-green, with yellow bar and white spot on wing; tail black-tipped with buff; eye-stripe dark and eyebrow white; rump and flanks yellow; underparts white, tinged with buff; legs blackish; bill black; eyes dark brown. Female very similar, but with upperparts streaked with dark and lightish brown.

THE TINY RIFLEMAN CREEPS IN A SPIRAL PATH UP *tree trunks, probing crevices in the bark for insects and grubs, and snatching prey from nearby leaves.*

Diet: insects and spiders.

Breeding: August to January. Builds domed woven nest of plant debris, lined with feathers, in a crevice, wall or hollow branch, up to 18 m (60 ft) above ground. 4–5 white eggs, incubated by both parents. Young hatch very underdeveloped, cared for by both parents. May have 2 broods a year.

Other information: lives in family groups. Unpaired males may assist breeding pairs. Young of first brood may help care for second.

SHARPBILL

Oxyruncus cristatus

Family: Oxyruncidae

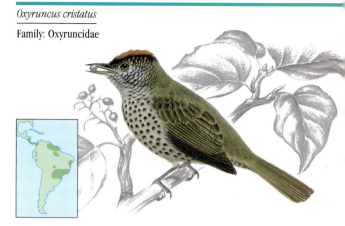

DISTRIBUTION: fragmented distribution in tropical rainforests and cloud forests of Central America and eastern South America, from Costa Rica to Paraguay, up to 1,800 m (6,000 ft).

SIZE: L 15 cm (6 in); WT 37–46 g (1.3–1.6 oz).

FORM: small, stocky bird with longish wings, short legs and conical, sharply-pointed bill. Color of upperparts olive-green, with dark brown wings and tail; crown has long, silky yellow-and-scarlet crest, bordered with black, usually hidden when not displaying; underparts white to pale yellow, with many dark brown spots; legs pale brown; bill dark brown; eyes light brown.

THE SHARPBILL IS A LITTLE-KNOWN BIRD THAT TRAVELS *rapidly through the tree tops in search of insects, spiders and berries. It often hangs upside-down to reach dangling fruits.*

DIET: insects, spiders; fruits, berries.

BREEDING: nest a shallow cup of loosely woven stems and dry leaves, coated with spider webs and plant material, glued with saliva to a high branch. Breeding habits only known from one nest; it is thought female feeds chicks on regurgitated fruit.

OTHER INFORMATION: song a slightly descending trill. The male has a serrated edge on his primary feathers, which may be used to produce sound, although not in display flights.

SUPERB LYREBIRD

Menura novaehollandiae

Family: Menuridae

DISTRIBUTION: dense forests and woodland, to above snowline, in southeastern Australia. Introduced to Tasmania.

SIZE: L 74–84 cm (29–33 in); WT 746 g (1.65 lb); male larger than female.

FORM: large pheasant-like bird. Color rich brown above, grayish-brown below; throat tinged rufous; bare facial skin bluish-black; crown slightly crested; legs and bill gray; eyes brown with pale eye-ring. Male has lyre-shaped, train-like tail.

A MALE LYREBIRD SHOWS OFF TO A FEMALE ON A LOW *display mound, fanning his shimmering silvery tail and turning slowly as he pours out a mellow stream of song.*

DIET: insects, grubs, worms.

BREEDING: June to September. Males polygamous. Makes large domed nest of sticks, leaves, roots, bark and moss, with side entrance, lined with feathers, on ground under cover. Single gray or purplish-brown egg, streaked and spotted, incubated by female for 35–40 days. Young hatch well-developed.

BARN SWALLOW

Hirundo rustica

Swallow
Family: Hirundinidae

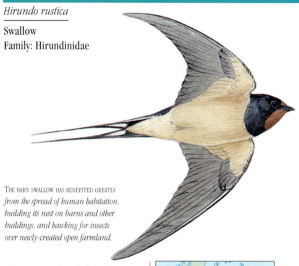

THE BARN SWALLOW HAS BENEFITED GREATLY *from the spread of human habitation, building its nest on barns and other buildings, and hawking for insects over newly-created open farmland.*

DISTRIBUTION: breeds throughout Europe, Asia and North America, excluding north, in open country near water; winters in southern hemisphere.

SIZE: L 18 cm (7 in); WT 11.0–24.8 g (0.4–0.9 oz).

FORM: has long, tapering wings, deeply-forked tail and short bill, with wide gape. Color of upperparts dark metallic blue; forehead chestnut; chin and throat dark chestnut separated from creamy to rufous-buff underparts by dark blue band; legs and bill black.

DIET: insects.

BREEDING: from May to September. Pair build deep cup-shaped nest of mud and straw, lined with feathers, on ledge on building or cliff, or in cave or open shed. 4–5 white eggs, spotted with red. Incubated mainly by female for 14–15 days. Young fledge at about 21 days.

OTHER INFORMATION: drinks on the wing, skimming the water surface.

Hirundo daurica

Family: Hirundinidae

THE RED-RUMPED SWALLOW *hunts in the air, catching insects in its wide mouth. It stops regularly to rest on telephone wires or thin branches.*

DISTRIBUTION: wide range of habitats, from southern Europe and North Africa east to southern Siberia, the Himalayas, Indian subcontinent, Korea and Japan, up to 4,700 m (15,500 ft).

SIZE: L 18 cm (7 in); WT 15–19 g (0.5–0.7 oz).

FORM: long, tapering wings and deeply- forked tail. Color of upperparts blue-black, with chestnut forehead, nape and rump; wings and tail metallic blue; underparts buff, streaked with brown; legs brownish-black; bill black; eyes dark brown.

DIET: flying insects.

BREEDING: season varies with location. Pair build closed bulbous nest of mud pellets mixed with grass and plant fibers, with long tubular entrance, attached to roof of cave, rock cleft or building. Nest thinly lined with feathers or wool. 3–5 white eggs, sometimes speckled with reddish-brown. Incubated by both parents for 14–15 days. Young cared for by both parents. Fledge at 23–25 days; roost in nest and are fed by parents for several more days.

OTHER INFORMATION: flight call a thin, shrill "quiitsch"; song chattering.

YELLOW WAGTAIL

Motacilla flava

Blue-headed wagtail, White-headed wagtail, Black-headed wagtail
Family: Motacillidae

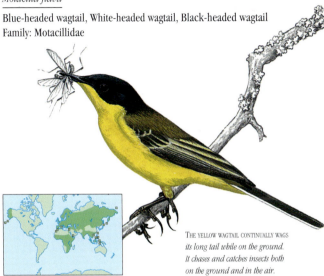

THE YELLOW WAGTAIL CONTINUALLY WAGS
*its long tail while on the ground.
It chases and catches insects both
on the ground and in the air.*

DISTRIBUTION: marshy grasslands and farmland across most of Europe and Asia to coastal Alaska. Winters in tropical Asia and Africa.

SIZE: L 17 cm (6.5 in); WT 11.2–27.0 g (0.4–0.95 oz).

FORM: slender bird with long tail and long toes. Male has greenish shoulders and wings, two pale wing-bars; tail blackish-brown with white edges; crown bright yellow, black, white or bluish-gray; legs black. Female duller.

DIET: mainly insects.

BREEDING: in spring and summer. Builds well-concealed cup-shaped nest of grasses, stems and roots, lined with hair, on ground. 5–6 pale buff or gray eggs, speckled with yellowish- or brownish-buff, incubated mainly by female for 12–14 days. Young cared for by both parents. Leave nest at 10–13 days; fledge at about 17 days.

OTHER INFORMATION: forms flocks on migration.

SKY LARK

Alauda arvensis

Skylark
Family: Alaudidae

THE SKY LARK IS FAMOUS FOR *its long, varied song, uttered in flight, often so high above the ground as to be almost invisible. It hovers in the air, still singing, then drops silently.*

DISTRIBUTION: open country and farmland from Europe and northernmost North Africa to Asia and Japan, except in the far north. Introduced to Australia, New Zealand, Japan and Vancouver Island, Canada.

SIZE: L 18 cm (7 in); WT 29–51 g (1–1.8 oz).

FORM: small ground-dwelling bird with long hind claws and longish tail. Color of upperparts brown, with darker streaks; wings darker, with white edges; tail edged in white; underparts buffish-white; head has short crest; legs and bill yellowish-brown; eyes brown.

DIET: mainly seeds and leaves.

BREEDING: from late spring. Nest a shallow cup made of dry grasses, lined with finer grasses, on ground. 3–4 greenish-gray or creamy-gray eggs, heavily mottled with dark grayish-brown, incubated by female for about 11 days. Young cared for by both parents. Leave nest at 9–10 days. Fledge at about 20 days.

OTHER INFORMATION: when alarmed, crouches close to ground and "freezes". Lives in flocks outside breeding season.

SHORE LARK

Eremophila alpestris

Horned lark
Family: Alaudidae

THE SHORE LARK IS AT HOME ON THE BEACH *or in the mountains, foraging in sand or snow. Unlike the Sky lark, it sings mainly on the ground.*

DISTRIBUTION: breeds on tundra, steppes, deserts and rocky alpine areas of Europe, Asia and most of North America, and in Atlas Mountains of North Africa; also small population around Bogota, in Colombia, South America. Winters on coast and arable land at higher altitudes.

SIZE: L 15 cm (6 in); WT 38–39 g (1.3–1.4 oz).

FORM: squat, short-legged lark. Color pinkish-brown with darker streaks; head patterned in black and yellow, with short black "horns"; female has less black on head; underparts whitish; legs black; bill grayish yellow.

DIET: seeds, buds, insects, mollusks, crustaceans.

BREEDING: from mid-May. Nest a loose cup of dry grass and stems, lined with plant down and hair, on ground. May build ring of peat or pebbles around nest. 2–7 pale greenish-white eggs, heavily speckled with yellowish-brown, incubated by female for 10–14 days. Young cared for by both parents. Leave nest at 9–12 days. May produce 2 broods each year.

OTHER INFORMATION: call a shrill "tsip" or "tseep"; also chirps; song similar to that of Sky lark. Flight swift, undulating.

125

RED-WHISKERED BULBUL

Pycnonotus jocosus

Family: Pycnonotidae

DISTRIBUTION: forests, plantations, clearings and gardens of India, Andaman Islands, Nepal, China and southeast Asia. Introduced to New South Wales, Australia, Nicobar Islands, Mauritius.

SIZE: L 20 cm (8 in); WT 25–31 g (0.9–1.1 oz).

FORM: smallish bird with long erect crest; wings short, rounded; legs short. Color of upperparts brown, darker on wings and tail; white cheek patches and small red patch behind eye; underparts white, with broken brown collar on breast; undertail bright crimson.

DIET: fruits, berries, buds, nectar; also insects, especially small beetles and ants.

THE RED-WHISKERED BULBUL IS A COMMON *garden bird with a cheerful whistling voice. It usually sings from the top of a bush, then flies to another song-post and starts again. A popular bird, it is often kept as a pet.*

BREEDING: builds cup-shaped nest of twigs, stems, grasses and other plant material, lined with finer grasses and rootlets, usually in bushes or small trees. 2–3 white or pinkish-white eggs, heavily marked with dark brown, black or gray, incubated by both parents for 12–14 days. Young cared for by both parents. Fledge at about 14–18 days, but are fed by both parents for several more days.

OTHER INFORMATION: lives in pairs or small flocks.

BLUE-BACKED FAIRY BLUEBIRD

Irena puella

Fairy bluebird
Family: Irenidae

As a fairy bluebird flies through the forest, his *iridescent plumage flashes blue in the shafts of sunlight. Flocks of bluebirds travel from one fruiting fig tree to another, calling.*

DISTRIBUTION: scattered populations in dense forests of southeast Asia, from western India and Nepal to Philippines.

SIZE: L 27 cm (10.5 in); WT 51–70.3 g (1.8–2.5 oz).

FORM: medium-sized bird with long, dense, fluffy plumage. Color of male: upperparts iridescent blue, with velvety-black cheeks, chin, wings and under-parts; under-tail coverts blue; legs and bill blackish; eyes red. Female dull bluish-green; primaries and tail dark brown; black patch around eye.

DIET: fruit, especially figs; also nectar and insects.

BREEDING: March to May. Builds flimsy cup nest of twigs, rootlets and mosses, lined with moss, in bush or small tree. 2 pale gray, reddish or buff eggs, marked with brown, purple and gray, incubated by female alone. Further details unknown.

OTHER INFORMATION: often sings whistling song from exposed perch; also has a two-note call "wit-weet". Lives in pairs or small flocks. Forages in undergrowth for berries and insects.

BLACK-FACED CUCKOO-SHRIKE

Coracina novaehollandiae

Large cuckoo-shrike
Family: Campephagidae

A BLACK-FACED CUCKOO-SHRIKE SITS *on a perch, scanning the foliage for insect prey. In size and shape it resembles a cuckoo, while its slightly hooked bill is like that of a shrike.*

DISTRIBUTION: breeds in forests and woodlands of India, southeast Asia, Australia and New Guinea; birds of southern populations winter farther north.

SIZE: L 33 cm (13 in); WT 87–102 g (3.1–3.6 oz).

FORM: medium-sized bird with long body, long, pointed wings and long, broad, rounded tail. Color of upperparts dove-gray; forehead, face and throat black; primaries black, edged with white; end of tail black, tipped with white; breast dark gray; rest of underparts white; legs and bill black; eyes dark brown.

DIET: insects, fruits, berries and seeds. Feeds mainly among foliage; also hovers over trees or grassland.

BREEDING: August to February. Pair builds shallow cup nest of twigs and bark bound with cobwebs, in high tree fork and camouflaged with bark and lichens. 2–3 olive, olive-brown or bluish-green eggs, spotted and blotched with brown, brownish-red and gray at larger end. May have 2 broods a year.

OTHER INFORMATION: has habit of repeatedly refolding wings after alighting on perch. Swoops on prey from perch. Sometimes feeds on ground.

BOHEMIAN WAXWING

Bombycilla garrulus

Waxwing
Family: Bombycillidae

THE RED BLOBS ON THE WING FEATHERS
*of the male Bohemian waxwing were
thought to resemble the red wax seals
once used on letters. It is a fast flier and
can capture insects on the wing.*

DISTRIBUTION: breeds in dense conifer-
ous or mixed forests in northern
Europe, Asia and North America.
Sometimes moves far south in winter.

SIZE: L 20 cm (8 in); WT 46.5–69 g
(1.6–2.4 oz).

FORM: plump bird with stiff, steeply-
inclined crest. Color of upperparts,
including crest, pinkish-buff, grayer on
back, shoulders and rump; end part of
tail black, tipped with yellow; primaries
black, tipped with black and yellow
bars; cheeks and forehead flushed
orange to deep pink; eye stripe and chin
black; underparts orange to pale wine
red, darker under tail; legs and bill
black; eyes dark brown. Female grayer.

DIET: berries; insects and larvae.

BREEDING: in late spring and summer.
Pair build cup nest of twigs, grasses
and lichens, lined with hair and down.
4–6 pale blue or grayish-blue eggs,
thinly spotted with black and gray,
incubated mainly by female for 13–14
days; male brings food.

OTHER INFORMATION: forms flocks in
winter.

PALMCHAT

Dulus dominicus

Family: Dulidae

PALMCHATS ARE NOISY, SOCIABLE *birds that perch very close together and roost in a huge communal nest. They feed high in the trees on flowers and berries.*

DISTRIBUTION: open woodland, plantations and pine clumps on Hispaniola and Gonave Island, up to 1,370 m (4,500 ft) in the West Indies.

SIZE: L 18 cm (7 in); WT 42 g (1.5 oz).

FORM: small bird with stout, pointed bill, short wings and long tail. Color of upperparts grayish-white to olive-brown, with darker head; wings yellowish-green; rump dark green; underparts yellowish-white, striped with dark brown; legs blackish; bill brownish; eyes brown.

DIET: berries and flowers, often feeding in large groups.

BREEDING: March to June. Breeds communally, as many as 30 pairs making large communal nest up to 1 m (3.3 ft) across, made of twigs, woven around trunk and leaves high in palm tree. Pairs have separate nest entrances and nest chambers, thinly lined with grass and pieces of bark. 2–4 white eggs, spotted with dark gray.

OTHER INFORMATION: has range of calls, both musical and harsh.

NORTH AMERICAN DIPPER

Cinclus mexicanus

American dipper, Mexican dipper
Family: Cinclidae

DISTRIBUTION: mountain streams from 600 m (2,000 ft) to tree line, from Alaska to southern Mexico.

SIZE: L 18–21 cm (7–8 in); WT 43–66 g (1.5–2.3 oz).

FORM: small, plump bird with short tail and long legs. Color grayish-brown, with white eye ring; legs yellowish; bill gray; eyes dark brown.

DIET: small fish and aquatic insects, especially caddis fly larvae.

BREEDING: May to June. Pair builds large domed nest of moss, lined with dead leaves, located in a steep bank or

THE NORTH AMERICAN DIPPER HUNTS UNDERWATER *for fish and aquatic insects. It can be seen on large stones in fast-flowing streams, constantly bobbing up and down ("dipping") as it scans the water.*

under a bridge over fast-flowing water, or under a waterfall; nest often used for several years. 4–6 white eggs, incubated by female for 13 days. Chicks cared for by both parents. Fledge at 18 days. Chicks can swim and dive expertly before they can fly

OTHER INFORMATION: has white nictitating membrane, which covers its eyes when underwater. When blinking, the flash of white also acts as a signal of alarm, aggression or excitement.

NORTHERN WREN

Troglodytes troglodytes

Winter wren, Common wren, Wren
Family: Troglodytidae

THE NORTHERN WREN IS A SMALL BIRD WITH *a big voice. Its exuberant song can be heard from low-growing bushes and briars from late winter until well into the summer.*

DISTRIBUTION: throughout temperate regions of northern hemisphere, from Europe and North Africa to Asia, Japan, Alaska, central and southern Canada, parts of eastern USA. Northern populations winter farther south.

SIZE: L 8 cm (3 in); WT 8.0–12.7 g (0.28–0.45 oz).

FORM: very small, plump bird with distinctive short, up-turned tail. Upperparts and wings reddish-brown, barred with darker brown; underparts buffish-brown; eyebrow pale; legs pinkish-brown; upper mandible blackish-brown, lower mandible pale yellowish-brown; eyes dark brown.

DIET: insects, spiders and other small invertebrates.

BREEDING: in spring and summer. Male makes several domed nests of leaves, grasses and mosses, in low-growing vegetation, rock or wall crevice or old bird nest. Female chooses one, lining it with feathers. 5–8 white eggs, sometimes with tiny spots or speckles of black or reddish-brown, incubated by female for 14–17 days. Chicks cared for by both parents. Young fledge at 15–20 days. Often has 2 broods a year.

OTHER INFORMATION: flight low, whirring.

NORTHERN MOCKINGBIRD

Mimus polyglottos

Mockingbird
Family: Mimidae

DISTRIBUTION: open country, deserts and urban areas of North America, from southern Canada to Mexico and Caribbean islands; introduced to Hawaii and Bermuda.

SIZE: L 23–28 cm (9–11 in); WT 36.2–55.7 g (1.3–2 oz).

FORM: has long, tapered tail. Color gray; wings black with two white wing bars; tail dark, with white outer feathers; legs and bill grayish; eyes light yellow.

DIET: fruits, berries, seeds and insects.

BREEDING: in spring and early summer. Pair builds cup nest of twigs, stems, other plant material and debris, lined

THE NORTHERN MOCKINGBIRD IS FAMED FOR *its ability to mimic (or "mock") the calls of other birds. It sings both by day and by moonlight; this habit has led to its nickname of American nightingale.*

with grasses, rootlets and hair or plant down, in low shrub. 3–5 pale blue or greenish-blue eggs, sometimes pinkish, spotted, speckled and blotched with various shades of red or reddish-brown and pale lilac; incubated by female for 11–14 days. Chicks cared for by both parents. Fledge at 12–14 days.

OTHER INFORMATION: as it runs along ground, it repeatedly raises and partly opens its wings. Song involves repetition of various phrases, some of them borrowed from other birds.

133

DUNNOCK

Prunella modularis

Hedge accentor, Hedge sparrow
Family: Prunellidae

ONE OF THE MOST POPULAR *songsters of hedgerows and gardens, the dunnock is a shy bird that seldom flies far. It creeps along the ground, probing among the dead leaves for insects and spiders.*

DISTRIBUTION: breeds in woodlands, parks, gardens, heaths and scrublands throughout Europe, Asia Minor, Iran and the Caucasus. Northern populations winter farther south as far as the Mediterranean and northern Africa.

SIZE: L 14 cm (5.5 in); WT 13–26 g (0.5–0.9 oz).

FORM: small inconspicuous bird. Color reddish-brown, streaked with black, darker on crown; head, breast and throat bluish-gray, except for brownish cheek patch; narrow buff wing-bar; legs pinkish-brown; bill blackish-brown; eyes brown.

DIET: insects, spiders, worms; seeds.

BREEDING: from early spring. Pair builds stout cup nest of twigs, stems, roots, dry leaves and moss, lined with moss, hair and wool. 3–6 bright blue eggs, incubated by female for 12–13 days. Chicks cared for by both parents; fledge at 12 days. May have 2 or 3 broods each year.

OTHER INFORMATION: sometimes makes up threesomes of 2 males and a female, which hold a territory. Extra male may also mate with female.

REED WARBLER

Acrocephalus scirpaceus

Family: Sylviidae

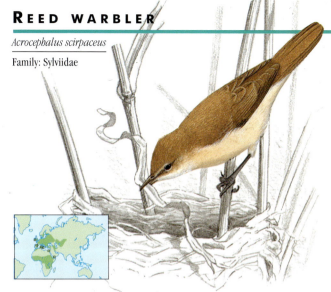

A FEMALE REED WARBLER WEAVES A NEST OF GRASSES and other plants fixed to the stems of reeds. The deep nest helps prevent the young falling out as the reeds are swayed by the wind.

DISTRIBUTION: waterside vegetation from Europe and northwest Africa to southwest Asia. Winters farther south in tropical Africa.

SIZE: L 13 cm (5 in); WT 8.0–19.7 g (0.3–0.7 oz).

FORM: color of upperparts brown, redder on back; cheeks tawny; narrow dark eyestripe, with pale eye-ring; wings darker, with pale tips; tail dark brown; underparts buffish, white on throat; legs grayish; bill yellowish; eyes brown.

DIET: insects, spiders, worms and other small invertebrates; berries.

BREEDING: in spring and summer. Nest a deep cup of woven grasses, stems, flower heads, leaves, reeds and spider webs, lined with finer material and feathers. 4 pale green or greenish-white eggs, heavily spotted, speckled and blotched with green, olive and gray at larger end; incubated by both parents for 11–12 days. Chicks cared for by both parents; fledge at 11–13 days.

OTHER INFORMATION: song an unmusical repeated "churr".

FIRECREST

Regulus ignicapillus

Family: Sylviidae

DISTRIBUTION: breeds in woodlands, hedgerows, parks and gardens from western and central Europe, North Africa, Madeira and Canary Islands to Asia Minor and Black Sea. Northern populations winter farther south.

SIZE: L 9 cm (3.5 in); WT 4.5–8.2 g (0.2–0.3 oz).

FORM: tiny warbler. Color of upperparts yellowish-green, with two whitish wing-bars; tail feathers white-tipped; head has bright orange-red crest, bordered by two black stripes that meet across forehead; throat and breast pale grayish-brown; belly white; legs brown; bill black; eyes brown. Female has lemon-yellow crest.

THE FIRECREST IS ONE OF EUROPE'S TINIEST BIRDS, *barely 9 cm (3.5 in) long. Although it is a warbler, it behaves rather like a tit, hanging upside down as it probes buds for insects.*

DIET: insects and their larvae, spiders, many of minute size.

BREEDING: starts in May. Pair build deep, thick hammock-like nest of mosses, lichens and spider webs, lined with feathers. 7–10 white to pale buff eggs, with fine pale brownish speckles, incubated by female for 14–17 days. Young cared for by both parents. Fledge at 16–21 days.

OTHER INFORMATION: call a high-pitched "tseet"; song a repeated high-pitched "siss".

AFRICAN PARADISE FLYCATCHER

Terpsiphone viridis

Paradise flycatcher
Family: Monarchidae

DISTRIBUTION: forests, riverside woodlands and green suburbs of Africa, south of Sahara.

SIZE: L 40 cm (16 in); WT 11.3–17.0 g (0.4–0.6 oz).

FORM: head has crest; male has streamer-like central tail feathers up to 20 cm (8 in) long. Male's upperparts and tail rich chestnut (white in some races); primaries brownish, edged with chestnut; head, neck and underparts metallic bluish-gray; undertail and underwing coverts white; legs slatey-gray; bill blackish; eyes brown, with blue eye-ring. Female less glossy.

DIET: insects.

A MALE AFRICAN PARADISE FLY-catcher with his latest prey. These birds make very small nests, so the male's tail can hang outside while incubating. The young soon outgrow this nest.

BREEDING: nonseasonal. Makes small bowl-shaped nest of plant fibers and rootlets, bound with spider webs, decorated with lichens, on open twig over water. 2–3 white or cream eggs, spotted with red and purplish-gray, incubated by both parents. Fledge at about 10 days.

OTHER INFORMATION: calls frequently, a sharp "zwee-zwer".

PIED FLYCATCHER

Ficedula hypoleuca

Family: Muscicapidae

DISTRIBUTION: forests, woodlands, parks and gardens, from western Europe and northwest Africa to western Asia. Winters in tropical Africa.

SIZE: L 13 cm (5 in); WT 9.7–14.3 g (0.3–0.5 oz).

FORM: color of male: upperparts black, blackish-brown or brown, often grayer on nape, shoulders and rump; broad white wingbar; outer tail feathers white; white splash or spots on forehead; underparts white; legs and bill black; eyes brown. Female quite similar, brown instead of black, with smaller wingbar, darker forehead and "dirty"-white underparts.

PIED FLYCATCHERS CATCH INSECTS IN THE AIR. THEY *wait on a perch until an insect comes within reach, then fly out, snatch it and return to another perch to eat it.*

DIET: insects, spiders; also worms and berries in the fall.

BREEDING: in summer. Female builds loose cup nest of grasses, leaves, roots and other plant material, lined with grass and hair, in tree hole, crevice in wall, or nestbox. 4–7 pale blue eggs, incubated by female for 12–13 days; male brings food. Chicks fledge at 13–16 days.

OTHER INFORMATION: when perching, regularly flicks wings and dips tail.

FLAME ROBIN

Petroica phoenicea

Bank robin, Robin redbreast
Family: Muscicapidae

THE FLAME ROBIN IS ACTUALLY A *fly-catcher, pouncing on insects flying past or chasing them through the air. The male has an attractive high-pitched tinkling song, said to sound like "you may come, if you will, to the sea".*

DISTRIBUTION: breeds in eucalypt forests and open woodlands of foothills of southeastern Australia and Tasmania; winter in more open low-land country.

SIZE: L 14 cm (5.5 in); WT 13.3 g (0.5 oz).

FORM: small plump-breasted bird. Color of male: upperparts dark gray, with small whitish patch above bill; white wingbar; tail edged in white; chin white; throat and underparts bright flame-red; lower belly area white. Female has buffish-brown upperparts with pale buff spot on forehead; pale buff bars on wings; underparts grayish-pale buff, darker on breast, sometimes tinged reddish-orange.

DIET: insects.

BREEDING: September to January. Builds bulky cup nest of strips of bark and grass bound with spider webs, decorated with moss, lichens or bark, lined with hair, fur or plant down, in tree hole, crevice in bark, rocks or buildings. 3–4 greenish-white eggs, spotted and speckled with reddish-brown and lilac, incubated for 12–14 days.

OTHER INFORMATION: lives alone or in large flocks outside breeding season.

WHITE-CRESTED LAUGHING THRUSH

Garrulax leucolophus

Family: Timaliidae

DISTRIBUTION: bamboo thickets, scrub and undergrowth of southeast Asia, from Himalayas to southwest China, up to 1,200 m (4,000 ft).

SIZE: l. 28 cm (11 in).

FORM: like a large thrush. Color chestnut, with paler belly and more olive-brown color on wings; tail blackish; head and breast white; head has erectable crest; black face mask; legs and bill blackish; eyes dark.

DIET: insects; berries, seeds.

BREEDING: builds well-hidden domed nest of dead leaves and mosses, with side entrance, in a shrub. 3–6 blue

THE WHITE-CRESTED LAUGHING-THRUSH LIVES IN *large flocks that from time to time burst into what has been described as "a chorus of diabolical cackling laughter". Birds appear to bounce over the ground as they forage for insects and berries.*

eggs, incubated for 13–16 days. Young fledge at 13–16 days.

OTHER INFORMATION: highly gregarious, forages in flocks of up to 100 birds; roosts in smaller groups. Has spectacular communal displays in which several birds dance together on forest floor, erecting crests and uttering a crescendo of laughing calls. At dawn and dusk other groups answer these calls. Forages among leaf litter on ground by digging, and by flicking over debris.

PIED BABBLER

Turdoides bicolor

Bicolored babbler
Family: Timaliidae

THE PIED BABBLER IS A SOCIABLE *bird, living in noisy groups of up to twelve birds. Its call is a high-pitched "skerr-skerr-skerr-kikikikikikerrkerrkerr...". One bird starts to call, or "babble", then all the others join in, getting shriller and shriller.*

DISTRIBUTION: dry scrubland and open woodlands of southwest Africa.

SIZE: L 25.4–26.7 cm (10–10.5 in); WT 77.3 g (2.7 oz).

FORM: medium-sized thrush-like bird with strong, slightly down-curved bill and strong legs and feet. Color white, with black wings, shoulders and tail; legs brownish-black; bill black; eyes reddish-orange.

DIET: insects and their larvae.

BREEDING: October to January. Nests communally in clump of thorn trees.

Builds cup nest of twigs and dry grasses, lined with hair, fine plant fibers and rootlets. 2–5 pale bluish-white eggs, incubated for 14–15 days. Young fledge at 13–16 days.

OTHER INFORMATION: each group defends a territory against other groups. Flocks often contain other birds. Usually keep to the cover of brush, creepers, shrubs and trees.

EASTERN BLUEBIRD

Sialia sialis

Family: Turdidae

THE EASTERN BLUEBIRD OFTEN HAS *a hunched appearance while perching. It may fly down from the perch to seize an insect on the ground. Suffers from competition for nest sites, but thrives in special nestboxes.*

DISTRIBUTION: breeds in open country with scattered trees, clearings, gardens and parks from east Canada through USA east of rockies to Gulf coast, western Mexico and Honduras. Northern populations winter farther south.

SIZE: L 14–19 cm (5.5–7.5 in); WT 31.6 g (1.1 oz).

FORM: small, plump bird. Male has deep blue upperparts, with chestnut throat, sides of neck, breast and flanks; wings tipped with black; belly white; legs dark brown; bill gray; eyes dark brown. Female similar but duller.

DIET: insects (especially crickets, grasshoppers, beetles, larvae); fruits.

BREEDING: pair builds loose cup nest of dry grasses, stems and small twigs, lined with grasses, in tree hole, hollow post, old nest or birdbox. 3–7 pale blue eggs, incubated by both parents, mainly female, for 12 days. Chicks cared for by both parents. Fledge at 15–18 days, but remain with male while female starts second brood. May produce several broods a year.

OTHER INFORMATION: when perched or flying, it utters a liquid "chur-lee".

EUROPEAN ROBIN

Erithacus rubecula

Robin, Robin redbreast
Family: Turdidae

DISTRIBUTION: woodlands, parks, gardens from western Europe and North Africa to western Siberia and Iran.

SIZE: L 14 cm (5.5 in); WT 14.2–22.5 g (0.5–0.8 oz).

FORM: small plump bird. Upperparts olive-brown; forehead, throat and breast orange-red, bordered by grayish-blue band; legs, bill and eyes brown.

DIET: insects, larvae, spiders and worms.

A EUROPEAN ROBIN DISPLAYS TO A FEMALE, SHOWING off his red breast. Both male and female robins defend territories all year round, singing a melodious warbling song from perches at strategic points.

BREEDING: in spring and summer. Female builds sturdy cup nest of dead leaves, grasses and moss, lined with hair and fine roots, in hollow log, tree trunk, stump, bank or low shrub. 5–6 white eggs, spotted, speckled and faintly blotched with light brown or pinkish-buff, incubated by female for 12–15 days; male brings food. Chicks cared for by both parents; fledge at 12–15 days. May have 2 or 3 broods a year.

OTHER INFORMATION: red breast used as warning signal to rivals.

SONG THRUSH

Turdus philomelos

Family: Turdidae

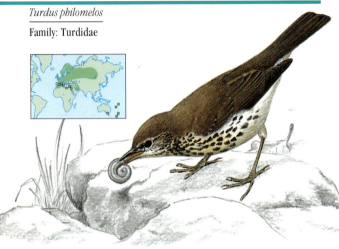

DISTRIBUTION: breeds in woodlands, copses, hedgerows and gardens from Scandinavia, western Europe and North Africa to western Siberia and Iraq.

SIZE: L 23 cm (9 in); WT 52–89 g (1.8–3.1 oz).

FORM: color of upperparts warm brown; chin whitish; underparts creamy-white, spotted with dark brown; buffish-orange tinge on breast and flanks; in flight, underwings golden-brown; legs flesh-colored; bill golden-brown, darker at tip; eyes light brown.

DIET: worms, snails, insects, especially caterpillars; fruits in the fall.

A SONG THRUSH HAMMERS A SNAIL AGAINST A STONE TO smash its shell. Each bird has its own favorite anvil, a stone or stretch of stony path usually covered in broken shells.

BREEDING: in spring and summer. Female builds cup nest of grasses, twigs, roots, dead leaves, mosses and lichens, lined with wood pulp or mud, in tree, shrub or building up to 3.7 m (12 ft) above ground. 4–6 light blue eggs, sparsely spotted, speckled or blotched with black or purplish-brown, incubated by female for 11–15 days. Chicks fledge at 12–16 days.

OTHER INFORMATION: sings for most of year, a loud song made up of repeated phrases, often mimicking other birds.

GOLDEN WHISTLER

Pachycephala pectoralis

Cuthroat, Golden-breasted/
White-throated thickhead,
Thunderbird, Whipbird,
Family: Pachycephalidae

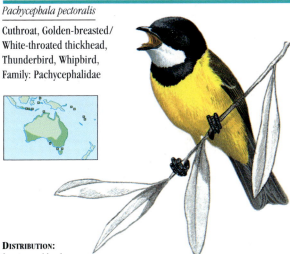

DISTRIBUTION:
forests, scrublands
and gardens of southern, eastern and
southwestern Australia, Tasmania, New
Guinea, Indonesia, Pacific islands.

SIZE: L 16 cm (6 in); WT 20.4–26.5 g
(0.7–0.9 oz).

FORM: color very variable: over 73 different island races. Upperparts of male
olive-green; tail black and gray; yellow
collar at nape joins rich-yellow underparts; black on head extends as broad
band between white throat and breast;
legs slatey gray; bill black; eyes red.
Female: upperparts grayish-brown
tinged with olive; throat mottled grayish-white; breast gray.

THE GOLDEN WHISTLER IS NAMED FOR ITS BRIGHT
*plumage and melodious, whistling song,
which ends with a whip-crack note.*

DIET: insects; also berries.

BREEDING: season variable. Pair build
shallow cup nest of grasses, leaves and
bark, bound with spider webs, lined
with grasses, in fork of small tree or
bush. 2–3 pale yellowish-buff eggs,
spotted with dark reddish-brown and
gray, incubated by both parents for
about 17 days. Chicks cared for by both
parents; fledge at 13–16 days.

OTHER INFORMATION: lives alone outside breeding season.

SPALDING'S LOGRUNNER

Orthonyx spaldingi

Northern logrunner, Black-headed logrunner,
Spalding's spinetail, Auctioneer-bird
Family: Orthonychidae

DISTRIBUTION: forest floor in dense
vegetation in rainforests of northeastern Australia.

SIZE: L 25–27 cm (9.8–10.7 in);
WT 126–213 g (4.4–7.5 oz).

FORM: plump bird with short legs and
spiny-tipped tail. Color of upperparts
and tail black; underparts white;
female has chestnut throat and breast;
legs gray; bill black; eyes brown.

DIET: insects; some berries.

BREEDING: May to August. Builds large,
bulky domed nest, loosely made of
twigs, bark, roots, grasses and mosses,
in thick vegetation up to 3 m (10 ft)
above ground, or on ground. Single
white egg, incubated for 17–21 days.

SPALDING'S LOGRUNNER IS A SECRETIVE FOREST BIRD
*that scratches rapidly on the ground for
insects, tossing leaves aside and using its
splayed-out spiny tail as a prop.*

Chicks hatch helpless; fledge at about
12–14 days.

OTHER INFORMATION: forms small
flocks. Rarely flies. When disturbed,
makes short, low whirring flight, then
lands and runs away into undergrowth.
Sings loudly at dawn and dusk, a
"chow chowchilla"; also makes raucous shouts and mimics other birds.
Have noisy territorial battles in spring.

VERDIN

Auriparus flaviceps

Family: Remizidae

DISTRIBUTION: dense scrubland and desert areas with thorn bushes and cacti, from the southern USA to northern Mexico.

SIZE: L 11 cm (4 in); WT 5.5–8.5 g (0.2–0.3 oz).

FORM: small bird with finely-pointed bill. Color dull brownish-gray, with chestnut shoulder-patches, and lemon yellow head and throat; female duller; legs and bill grayish; eyes dark brown.

DIET: insects and berries.

BREEDING: from early spring. Nests in fork of branch in shrub, low tree or

THE VERDIN IS A SMALL BIRD OF ARID AREAS, ABLE TO go for long periods without water. It flits to and fro between bushes, often hanging upside down as it probes for insects.

cactus. Nest a purse-like prickly mass of thorny twigs, lined with spider webs, grasses, dead leaves, feathers and plant down, with side entrance. 3–6 bluish-green eggs, finely speckled with reddish-brown, incubated by female or by both parents for 14 days. Chicks cared for by both parents. Fledge at about 3 weeks, but continue to roost in nest for some time.

OTHER INFORMATION: solitary outside breeding season. Individuals make their own nests for roosting in.

BLUE TIT

Parus caeruleus

Family: **Paridae**

THE BLUE TIT IS A SMALL, INQUISITIVE *bird, one of the first to learn new tricks at the bird-table. It is very agile, able to hang upside down from the thinnest of twigs as it feeds on buds and insects.*

DISTRIBUTION: woodlands, parks, gardens, from southern Scandinavia, western Europe, North Africa and Canary Islands to Volga river and Middle East.

SIZE: L 11 cm (4 in); WT 9–14 g (0.3–0.5 oz).

FORM: small plump bird with longish tail. Upperparts grayish-blue; shoulders yellowish-green; underparts sulfur-yellow; crown cobalt blue; forehead, cheeks and nape patch white; eyestripe and collar on nape black, linking with black bib; some races have darker head; legs dark slatey-blue; bill black; eyes dark brown.

DIET: insects and larvae; also buds, fruits, nuts and seeds.

BREEDING: in spring and early summer. Female builds cup nest of grasses, dead leaves, mosses, hair and spider webs, lined with hair and feathers, in hole in tree or wall, or in nestbox. 7–12 white eggs, finely speckled, spotted or blotched with purplish-red or reddish-brown, incubated by female for 12–16 days; male brings food. Chicks cared for by both parents; fledge at 15–23 days. May have 2 broods a year.

OTHER INFORMATION: forages in flocks, often with other tit species.

BOREAL CHICKADEE

Parus hudsonicus

Family: Paridae

THE BOREAL CHICKADEE IS A HARDY LITTLE BIRD —
it can survive winter temperatures as low as
−45°C (−50°F). Large flocks scour the winter
countryside for food.

DISTRIBUTION: coniferous and mixed
forests, often near bogs, in northern
North America, excluding shores of
Arctic Ocean, south to Canada/US border. May wander farther south in winter.

SIZE: L 14 cm (5.5 in); WT 7.0–12.4 g
(0.3–0.4 oz).

FORM: small bird with short wings and
short, sharp bill. Upperparts grayish-
brown, browner on head, with white
cheeks; underparts whitish, with black
bib and brown flanks; legs brownish-
gray; bill and eyes dark brown.

DIET: insects and larvae; also fruits,
berries, seeds.

BREEDING: in early summer. Makes
cup nest of hair and fur on platform of
moss, lichens and bark in a tree hole.
4–9 white eggs, finely spotted or speckled with light red, reddish-brown or
brown; incubated mainly by female for
about 12 days; male brings food. Chicks
cared for by both parents.

OTHER INFORMATION: call a nasal
"tseek-a-day-day".

LONG-TAILED TIT

Aegithalos caudatus

Family: Aegithalidae

THE LONG-TAILED TIT *seldom spends time on the ground. Small parties flit from tree to tree in bobbing flight, making penetrating "zee-zee-zee" calls.*

DISTRIBUTION: woodlands, scrublands and parks from Scandinavia and western Europe across Asia to Japan up to 2,740 m (9,000 ft). Northern populations may winter farther south.

SIZE: L 14 cm (5.5 in); WT 7.8–9.5 g (0.3 oz).

FORM: small bird with very long tail and very small bill. Upperparts black or gray; primaries blackish-brown, inner ones edged with white; tail blackish or dark gray; blackish triangle from shoulders to just above tail; underparts white, flanks pinkish; head may be pure white or have black or grayish eyestripe; legs blackish-brown; bill black; eyes brown.

DIET: mainly insects.

BREEDING: starts March to April. Pair build oval domed nest of moss bound with spider webs and hair, coated with lichens and lined with feathers, beside tree trunk or in large fork, up to 20 m (70 ft) above ground. 8–12 white eggs, sometimes finely speckled with purplish-red, incubated mainly by female for 12–14 days; male brings food. Chicks cared for by both parents; fledge at 14–18 days.

OTHER INFORMATION: lives and roosts in flocks.

EURASIAN NUTHATCH

Sitta europaea

European nuthatch, Nuthatch
Family: Sittidae

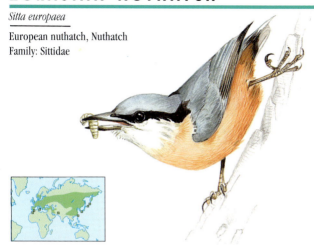

DISTRIBUTION: woodlands, parks and gardens from western Europe to Kamchatka (Russia) and Japan.

SIZE: L 11–13 cm (4–5 in); WT 19.5–24.0 g (0.7–0.8 oz).

FORM: small bird with short, square tail and strong, pointed bill. Color of upperparts bluish-gray, with black eye-stripe extending to side of neck; underparts range from white in northern populations to deep chestnut-yellow in southern birds; throat white; flanks chestnut; legs yellowish-brown; bill slatey-gray; eyes dark grayish-brown.

DIET: mainly insects; also seeds, nuts.

THE EURASIAN NUTHATCH CREEPS UP AND DOWN *tree trunks, searching for insects in bark crevices. Although narrow enough to probe the cracks, its bill is also strong enough to break open nuts.*

BREEDING: April and May. Female builds loose cup nest of dead leaves and pieces of bark in tree hole, crevice in wall or nestbox. Pair may plaster mud around large hole to make entrance smaller. 6–9 white eggs, spotted, speckled and blotched with shades of red, brown and purple; incubated by female for 14–18 days; male brings food. Chicks fledge at 23–25 days.

OTHER INFORMATION: unlike most birds, it climbs down trees head-first.

BROWN TREECREEPER

Climacteris picumnus

Family: Climacteridae

DISTRIBUTION:
waterside trees, dry forests and woodlands of eastern Australia.

SIZE: L 15 cm (6 in); WT 37 g (1.3 oz).

FORM: small bird with short legs and long, thin, downward-curving bill. Upperparts grayish-brown, darker on crown; eyebrow pale buff; eyestripe dark; tail grayish-brown with broad black band; throat creamy-white, with narrow black band; underparts buffish-brown, lower parts heavily streaked with black and cream; underside of tail barred black and cream; legs dark gray; bill black; eyes dark brown. Female has chestnut throat band.

THE BROWN TREECREEPER IS A SMALL, INCONSPICUOUS *bird that hops in a spiral up tree trunks in search of insects in the bark. It cannot climb down trees, so its flits to the bottom of the next tree and starts climbing again.*

DIET: insects, especially ants, other small invertebrates; seeds.

BREEDING: from June to January. Builds cup nest of grasses, hair and fur, lined with feathers, low in hollow tree, stump or post. 3–4 pinkish-white eggs, heavily spotted with dark red, incubated by both parents for 16–23 days. Young cared for by both parents. Fledge at 25–26 days.

OTHER INFORMATION: feeds on ground as well as in trees. Lives in small parties outside breeding season.

SCARLET-BACKED FLOWERPECKER

Dicaeum cruentatum

Family: Dicaeidae

DISTRIBUTION: lowland rainforests, secondary forests and mangrove swamps of northern India, southern China, Malay Peninsula, Sumatra, Borneo and Philippines, up to 1,200 m (4,000 ft).

SIZE: L 10 cm (4 in); WT 7–8 g (0.2–0.3 oz).

FORM: small, compact bird with short tail and short bill. Color of male: upperparts red; wings, tail and flanks black; underparts whitish; legs and bill dark gray; eyes dark. Female brown above, with darker wings and tail, and red rump.

DIET: mainly berries, especially mistletoe; also insects, spiders and nectar.

BREEDING: nest purse-shaped with side entrance, made by female of mosses,

THE SCARLET-BACKED FLOWERPECKER IS AN IMPORTANT *dispersal agent of mistletoes. It feeds on the berries, whose seeds pass right through it. As it wipes its vent on a branch to remove the seeds, they fall into the bark, where they germinate.*

plant down and spider webs, decorated with spider cocoons, seeds and insect excrement, suspended from leafy twig. 2 white eggs, incubated by female alone for 12 days. Chicks fledge at about 15 days.

OTHER INFORMATION: forms small flocks outside breeding season.

REGAL SUNBIRD

Nectarinia regia

Family: Nectariniidae

THE REGAL SUNBIRD FEEDS MAINLY ON NECTAR, *and has a slender, curved bill for reaching deep into flowers. Its tongue is tubular and acts as a straw through which it sucks up the nectar.*

DISTRIBUTION: mountain forests of western Uganda and western Tanzania.

SIZE: L 11 cm (4 in); WT 6 g (0.2 oz).

FORM: small bird with slender, down-curving bill. Male has brilliant irides-cent plumage, with olive upperparts and tail, greener on head; underparts bright red; flanks bright yellow; legs and bill blackish; eyes dark. Female dull, with olive-green upperparts and bright yellow underparts.

DIET: nectar, insects and spiders.

BREEDING: season varies according to timing of rainfall. Female builds bag-like nest with side entrance, suspended from a twig, made of grasses, lichens and other plant material, bound together with spider webs. 2–3 whitish or bluish eggs, incubated by the female for 13–15 days. Chicks are cared for by both parents and fledge at about 14–19 days.

OTHER INFORMATION: has a rapid trilling or warbling song.

CHESTNUT-FLANKED WHITE-EYE

Zosterops erythropleura

Family: Zosteropidae

DISTRIBUTION: forests and open woodlands of the Russian far east, Manchuria, North Korea and southern China. Northern populations winter farther south.

SIZE: L 11 cm (4 in); WT 9.9–11.5 g (0.3–0.4 oz).

FORM: very small bird with rounded wings and short legs. Upperparts, wings and tail yellowish-green; throat and upper breast yellow; belly and eye-ring white; flanks usually chestnut.

DIET: insects, nectar, berries, fruit.

BREEDING: in summer. Pairs for life. Builds cup nest in well-concealed tree fork. 2–6 whitish or pale blue eggs,

THE CHESTNUT-FLANKED WHITE-EYE IS SOMETIMES *kept as a cage bird for its attractive song. In the wild, it forages in flocks for fruits and insects, probing into crevices with its narrow, pointed beak.*

incubated by both parents for 11–12 days. Chicks cared for by both parents; fledge at 11 days.

OTHER INFORMATION: call a loud "tsee-plee". Uses high-pitched plaintive note to keep in contact over long distances; threatens other birds by clattering beak.

NORTHERN CARDINAL

Cardinalis cardinalis

Common cardinal,
Cardinal
Family: Emberizidae

DISTRIBUTION: woodland edges, copses, parks and gardens of North and Central America, from southeastern Canada to Belize; introduced to Hawaii and Bermuda.

SIZE: L 22 cm (8.5 in); WT 33.6–64.9 g (1.2–2.3 oz).

FORM: plump bird with heavy bill and tall crest on head. Color of male red, with black face and chin. Legs brown; bill pinkish; eyes brown. Female has greenish-brown upperparts, with red on wings, tail and crest; underparts pinkish-brown; face and chin fainter.

DIET: insects, seeds, fruit, flowers, buds, sap (left over by sapsuckers).

THE NORTHERN CARDINAL IS A HANDSOME BIRD THAT *often visits bird tables. Its loud, liquid whistling song can be heard all year round.*

BREEDING: in spring. Female builds cup nest of twigs, stems, grasses, bark, rootlets and vines mixed with other plant material and debris; lined with fine grasses, rootlets, lichens or hair. 2–5 white or greenish-white eggs, spotted, speckled and blotched with brown and pale purple or gray; incubated by female for 11–13 days; males feeds her. Chicks cared for by both parents. Leave nest at 9–11 days; fledge by 19 days; independent at 38–45 days. May have up to 4 broods a year.

OTHER INFORMATION: call a sharp sounding "chink".

PAINTED BUNTING

Passerina ciris

Family: Emberizidae

DESPITE HIS BRIGHT COLORS, THE MALE PAINTED *bunting can be surprisingly hard to see, as the blocks of color break up his outline. In spring, he advertises his presence by singing from a conspicuous perch.*

DISTRIBUTION: woodland edges, thickets, hedgerows, farmland and gardens of southern USA. Winters from Gulf coast to Panama and Cuba.

SIZE: L 12.5–14 cm (5–5.5 in); WT 12.9–19.0 g (0.5–0.7 oz).

FORM: small bird with short, heavy bill. Male bright red, with darker wings and tail; head, neck and cheeks violet-blue; shoulders and back glossy green; legs dark brown; bill yellowish-brown; eyes dark brown, with red eye-ring. Female has green upperparts, dusky yellow to amber underparts and yellow eye-ring.

DIET: mainly grass seeds; also insects.

BREEDING: in spring and early summer. Female builds deep cup nest of grasses, stems and leaves, lined with hair and grasses, woven around supports in bush, vine or tangle of Spanish moss, high in tree. 3–5 white eggs, finely speckled with chestnut-red and purple; incubated by female for 11–12 days. Young cared for by both parents. Fledge at 8–9 days. May then be fed by male while female starts next brood. May have up to 4 broods a year.

OTHER INFORMATION: shy, staying in cover outside breeding season.

SNOW BUNTING

Plectrophenax nivalis

Family: Emberizidae

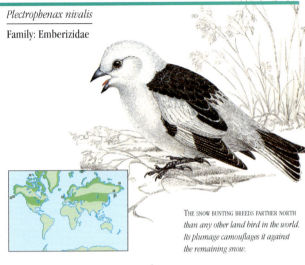

THE SNOW BUNTING BREEDS FARTHER NORTH *than any other land bird in the world. Its plumage camouflages it against the remaining snow.*

DISTRIBUTION: breeds on tundra and mountain tops throughout the Arctic; winters in coastal areas, farmlands and moorlands as far south as northern France, central Asia, California, Kansas and Virginia, USA.

SIZE: L 16.5 cm (6.5 in); WT 34–56 g (1.2–2 oz).

FORM: color of male in summer pure white, with black back, primaries and central tail feathers, with white wing-bar; in winter has reddish-brown head, cheeks, breast and rump, and mottled dark brown-and-white shoulders; legs black; bill black in summer, brownish-yellow in winter; eyes brown. Female like winter male, but is more brownish.

DIET: seeds and insects.

BREEDING: in summer. Pair build cup nest of mosses, grasses, stems and lichens lined with grasses, hair and feathers, on ground, in rock crevice or wall. 5–6 pale green, buff or gray eggs, heavily mottled, incubated mainly by female for 10–14 days. Chicks cared for by both parents. Leave nest at 8–10 days; fledge a few days later.

OTHER INFORMATION: tunnels into snow to keep warm in cold weather.

WHITE-THROATED SPARROW

Zonotrichia albicollis

Family: Emberizidae

DISTRIBUTION: breeds in open woodlands, forest edges and shrublands from Canada east of the Rockies to Pennsylvania, Wisconsin and Montana. Winters farther south, as far as the Gulf coast and northern Mexico.

SIZE: L 17 cm (6.25 in); WT 21.0–38.5 g (0.7–1.4 oz).

FORM: color of upperparts barred rusty-brown, with 2 white wingbars; crown has black and yellowish-white stripes; breast grayish, fading to white on belly; throat white, with narrow black mustache; flanks buff; legs pinkish-brown; bill yellowish-brown; eyes brown.

DIET: mainly insects and other invertebrates; also fruits and seeds.

THE WHISTLING OF THE WHITE-THROATED SPARROW *can be heard all year round. Like many sparrows, it feeds on the ground, scratching among the leaves for seeds and insects.*

BREEDING: in summer. Female builds cup nest of grasses, twigs and rootlets, lined with grasses and hair, on ground or in clump of vegetation. 4–6 pale blue or greenish-blue eggs, spotted, speckled or blotched with purplish- or chestnut-red and pale lilac; incubated by female for 11–14 days. Chicks cared for by both parents. Leave nest at 7–12 days; fledge a few days later.

OTHER INFORMATION: song a thin whistle, of two single notes, followed by 3 triple notes, said to sound like "Old Sam Peabody Peabody Peabody". Also makes loud "pink" call.

RED-LEGGED HONEYCREEPER

Cyanerpes cyaneus

Blue honeycreeper.
Family: Emberizidae

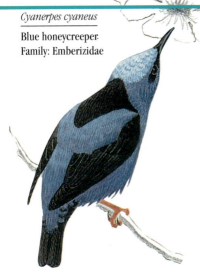

A RED-LEGGED HONEYCREEPER FEEDS ON nectar. *Its long, down-curved bill is well suited to probing trumpet-shaped flowers, and to probing in crevices for insects. Sometimes it cheats and bores a hole at the base of a flower to get to the nectar more easily.*

.DISTRIBUTION: woodlands and forest edges of Central America, including Cuba, and northern South America, as far south as Bolivia and Brazil.

SIZE: L 13 cm (5 in); WT 11.0–18.3 g (0.4–0.65 oz).

FORM: small bird with long, down-curved bill. Color of breeding male bright "electric" blue, with black shoulders and wings; wingbar blue; crown bright turquoise; black eye-patch; legs bright red; bill black; eyes dark. Female (and male outside the breeding season) light olive-green with more olive-brown wings; yellowish-buff eyestripe; underparts streaked green-and-white.

DIET: fruits, nectar, insects. Feeds among foliage. May leap into the air to catch flying insects.

BREEDING: female builds open cup nest in tree or shrub. 2 pale eggs, spotted with dark brown, incubated by female for 12–18 days. Chicks cared for by both parents, but mainly by female. Fledge at 13–14 days.

OTHER INFORMATION: lives in flocks.

MAUI PARROTBILL

Pseudonestor xanthophrys

Family: Fringillidae

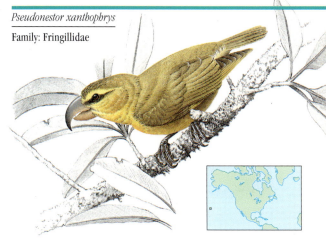

DISTRIBUTION: mountain rainforests with dense undergrowth above 1,500 m (5,000 ft) on the island of Maui, Hawaii.

SIZE: L 14 cm (5.5 in); WT 20 g (0.7 oz).

FORM: small finch-like bird with parrot-like bill. Color of upperparts yellow to olive-green, with darker wings and tail; yellow of underparts extends to cheeks, except for dark eyestripe; legs blackish; lower mandible flesh-colored; upper mandible dark gray on top, flesh-colored toward edges; eyes dark brown.

DIET: insects, especially larvae.

BREEDING: pair build open cup nest. 2–3 whitish eggs, scrawled with reddish-

THE MAUI PARROTBILL IS DESCENDED FROM A FINCH-*like species that colonized Hawaii many years ago. It has evolved a parrot-like bill, which it uses to chisel into branches to get at wood-boring grubs inside.*

brown or gray, incubated by female for 13–14 days. Chicks cared for by both parents, mainly female. Fledge at 15–22 days.

OTHER INFORMATION: song a descending series of notes. Calls include loud "kzeet" and thin "queet".

CONSERVATION STATUS: highly vulnerable. It occurs only on the island of Maui. Its habitat is greatly at risk from introduced animals such as pigs, goats and cattle.

EASTERN MEADOW-LARK

Sturnella magna

Family: Icteridae

DISTRIBUTION: grasslands and farmland, from eastern and southern USA to northern South America east of Andes. Northern birds fly south in winter.

SIZE: L 23–25 cm (9–10 in); WT 76–102 g (2.7–3.6 oz).

FORM: male has light brown upperparts, barred and streaked with darker brown; head has black-and-whitish stripes, short yellow eyebrow, broad white mustache; chin and underparts bright yellow, with broad black breast collar; undertail coverts whitish; legs pinkish-brown; bill light gray; eyes dark brown. Female (and male outside breeding season) similar, but paler.

A MALE EASTERN MEADOW-LARK SHOWS OFF THE BOLD *black-and-yellow markings on his breast to a female, who is also adopting a head-up, wings-down display posture.*

DIET: insects, seeds; feeds on ground.

BREEDING: in spring and summer. Female builds large, domed nest of grasses woven around nearby vegetation, with side entrance, lined with grasses. 3–5 white eggs, spotted, speckled or blotched; incubated by female for 13–15 days. Chicks cared for by both parents, mainly by female. Fledge at 11–12 days.

OTHER INFORMATION: has clear, mellow, whistling song,.

SOLITARY VIREO

Vireolanius solitarius

Blue-headed vireo
Family: Vireonidae

DISTRIBUTION: breeds in forests and other clumps of trees from Canada to Honduras, El Salvador and Nicaragua, and on Cuba. Northern populations winter farther south.

SIZE: L 14 cm (5.5 in).

FORM: color of upperparts light olive-green; wings and tail black with creamy bars; 2 white or yellowish wing-bars; head bluish-gray, with broad white ring that extends as stripe across base of bill; underparts white; flanks pale olive-green, streaked darker; legs and bill blackish-gray. Western race generally paler.

DIET: insects, berries.

THE SOLITARY VIREO IS NOT AT ALL EASY TO SEE, *as it feeds high in the trees and moves around rather slowly. However its rich song of "chu-wee-cheerio" readily gives away its presence.*

BREEDING: in spring and summer. Makes cup nest of bark, rootlets, grasses, lichens, mosses, feathers, plant down and debris, bound together with spider webs, attached to twigs around rim; lined with grasses, mosses, hair or fur. 3–5 white or creamy eggs, sparsely spotted or speckled with brown, chestnut or black, incubated by both parents, probably for 10–11 days. Chicks fledge at 11–13 days.

OTHER INFORMATION: either lives alone or in pairs.

BOAT-TAILED GRACKLE

Quiscalus major

Family: Icteridae

DISTRIBUTION: swamps of eastern and southern coasts of the United States.

SIZE: L 38–42 cm (15–18 in); WT 132–253 g (4.6–8.9 oz); male larger than female.

FORM: large bird with very long, keel-shaped tail. Male iridescent black, with purple sheen on back, head and underparts; legs black; bill blackish; eyes yellow to brown. Female tawny-brown; darker wings, tail and eye-patch.

DIET: worms and other invertebrates; vertebrates such as fish, frogs, eggs, young birds, mice; also seeds.

THE BOAT-TAILED GRACKLE GETS ITS NAME FROM *its long tail, shaped like the keel of a boat. Grackles are noisy gregarious birds that live and nest in flocks.*

BREEDING: season variable. Nests in colonies. Female builds large cup nest of swamp plants and grasses, bound together with decaying waterweeds and mud, lined with grasses and rootlets, in tree or bush; male stands guard. 3–4 pale blue, pinkish or purplish eggs, spotted, blotched, scrawled or scribbled with black, purple or gray; incubated by female for 13–14 days. Chicks cared for by female. Fledge at 20–23 days. May have 2 or 3 broods a year.

OTHER INFORMATION: has a variety of calls, including chucks, squeaks and rattles.

MONTEZUMA OROPENDOLA

Psarocolius montezuma

Family: Icteridae

THE MALE MONTEZUMA OROPENDOLA HAS A BIZARRE song, made up of gurgling and hollow popping noises. He also has a spectacular courtship display.

DISTRIBUTION: moist evergreen forests of Caribbean slopes of Central America.

SIZE: L 38–50.5 cm (15–20 in); WT 198–528 g (7–18.6 oz); male larger than female.

FORM: large bird with long tail, long, stout, pointed bill and shield-like extension of bill horn toward forehead. Color chestnut, brighter on upper wings, with blackish head, primaries and central tail feathers; rest of tail yellow; male has thin crest; short, pale pink mustache; pale blue patch of bare skin below eye; legs blackish; bill (and shield) bluish-black, with orange tip; light yellowish-brown eyes.

DIET: mainly fruits and nectar.

BREEDING: nests in colonies, up to 140 nests in same tree. Nest a woven bag of plant fibers, up to 180 cm (72 in) deep, slung from high branch of tree. 2 whitish eggs, incubated by female for 12–15 days. Chicks fledge in 37 days.

OTHER INFORMATION: may form flocks of over 100 birds.

AMERICAN GOLDFINCH

Carduelis tristis

Family: Fringillidae

LARGE, TWITTERING FLOCKS OF AMERICAN *goldfinches tour the countryside in winter, in search of weed seeds.*

DISTRIBUTION: open woodlands, roadsides, weedy fields of North America, from southern Canada to Mexico.

SIZE: L 11 cm (4 in); WT 8.6–20.7 g (0.3–0.7 oz).

FORM: small bird with short, stout bill. Color of male bright yellow, with black cap, black wings, with feathers edged with white; white wingbar; tail black-and-white; rump and undertail coverts white; legs pinkish-brown; bill yellowish-brown; eyes brown; non-breeding male brownish, with no dark cap. Female has dull olive back and head, yellow underparts; brown with grayish-brown underparts in winter.

DIET: seeds, buds, insects.

BREEDING: in spring and summer. Nest a cup of plant fibers, bark and wool, bound with spider webs and lined with plant down, well off ground in tree or shrub. 4–6 pale blue or greenish-blue eggs, incubated by female for 12–14 days. Chicks fledge at 11–17 days.

OTHER INFORMATION: nicknamed the "wild canary" for canary-like song.

RED CROSSBILL

Loxia curvirostra

Common crossbill
Family: Fringillidae

DISTRIBUTION: coniferous forests from northern North America, Europe and Asia, south to Guatemala, North Africa and the Himalayas.

SIZE: L 16 cm (6 in); WT 29.2–44.9 g (1.0–1.6 oz).

FORM: small bird with short, forked tail and stout bill, crossed at tip. Male brick red, brightest on crown and rump; wings, tail, legs, bill, eyes dark brown. Female yellowish-green, streaked on upperparts, brighter on upper breast and rump.

DIET: mainly conifer seeds.

BREEDING: season variable. Pair build

THE RED CROSSBILL'S STRANGE BILL IS ADAPTED FOR *separating the scales of pine cones and extracting the seeds.*

cup nest on platform of pine twigs, high in conifer; nest of mosses, grasses, lichens and wool, lined with grasses, hair, fur and feathers. 3–4 pale blue or bluish-white eggs, sparsely spotted, incubated by female for 13–16 days. Chicks fledge at 17–22 days.

OTHER INFORMATION: forage in flocks.

PINE GROSBEAK

Pinicola enucleator

Family: Fringillidae

THE PINE GROSBEAK FEEDS BOTH IN THE TREE TOPS *and on the ground. Its large, rounded bill is adapted for eating buds and berries.*

DISTRIBUTION: coniferous forests, woodlands and scrublands of northern and western North America, and from northern Scandinavia to Siberia.

SIZE: L 20 cm (8 in); WT 52–62 g (1.8–2.2 oz).

FORM: small, plump bird with short, forked tail and short, stout, hooked bill. Color of male red, with blackish wings and tail; white wingbars; grayish eye-stripe; lower belly white; legs, bill and eyes dark brown. Female similar, but greenish-yellow, with grayish shoulders and rump.

DIET: berries, buds.

BREEDING: in early summer. Female builds loose cup nest of twigs, lined with rootlets, grasses and mosses. 3–5 light blue eggs, sparsely spotted and blotched with black and purplish-brown or pale lilac, incubated by female for 13–14 days; male brings food. Chicks cared for by both parents. Probably fledge at about 13–14 days.

OTHER INFORMATION: in flight, utters whistling "pui-pui-pui" call. Flocks sometimes migrate considerable distances in search of food in winter.

HAWFINCH

Coccothraustes coccothraustes

Family: Fringillidae

DISTRIBUTION: broadleaved and mixed forests and parks across Europe, Asia and North Africa. Northern populations move south in winter.

SIZE: L 18 cm (7 in); WT 48–62 g (1.7–2.2 oz).

FORM: smallish bird with large head, thick neck and very large, stout, almost conical bill. Color of male "dirty" chestnut-brown, with black wingtips and white wing patch; tail has white tip; nape of neck gray; black bib extending to eye; lower belly white; legs light brown; bill grayish-blue in summer, brownish-yellow in winter; eyes reddish-brown. Female duller.

THE HAWFINCH'S HUGE BILL IS USED TO OPEN HARD *pitted/stoned fruits to extract the seeds, and to crack open the seeds. Its powerful jaw muscles make the bird's cheeks look fat.*

DIET: fruits, seeds, buds, insects.

BREEDING: from late spring. Pair build cup nest of twigs, roots and lichens, lined with plant fibers, rootlets and hair, in tree. 2–7 light blue or grayish-green eggs, spotted, scrawled and scribbled with blackish-brown; incubated, usually by female, for 9–14 days. Chicks fledge at 10–14 days.

OTHER INFORMATION: bill can exert a force of over 45 kg (100 lb) to crack open certain pits/stones.

EURASIAN GOLDFINCH

Carduelis carduelis

Goldfinch
Family: Fringillidae

DISTRIBUTION: open woodland, scrubland and gardens from North Africa and western Europe to the Middle East and central Asia. Introduced to Australia and New Zealand.

SIZE: L 12 cm (5 in); WT 13.4–17.8 g (0.5–0.6 oz).

FORM: color of upperparts tawnybrown; wings black with broad gold bar, white tips to primaries; tail black with white tip; crown, nape and sides of neck black; broad white band behind each eye; forehead, face and upper chin bright crimson; rump and underparts white; legs pale flesh; bill pinkish-white.

DIET: seeds, insects.

THE EURASIAN GOLDFINCH IS STILL WIDELY KEPT AS A *caged bird for its liquid, tinkling song, sung almost all day long in summer.*

BREEDING: spring to summer. Female builds cup nest of mosses, grasses, lichens, roots and plant down, lined with plant down and wool. 4–6 pale blue eggs, finely spotted, speckled, streaked and blotched with purple, purplish-black, pink or red at larger end; incubated by female for 12–14 days; male brings food. Chicks cared for by both parents. Fledge at 12–15 days; independent 1 week later. May have 2 or 3 broods a year.

OTHER INFORMATION: male performs dancing display flight, showing off his gold wingbars.

ZEBRA FINCH

Poephila guttata

Chestnut-eared finch
Family: Estrildidae

DISTRIBUTION: dry grasslands and other open country near water throughout most of Australia, excluding the far east; also on the Lesser Sunda islands.

SIZE: L 10 cm (4 in); WT 19 g (0.7 oz).

FORM: small bird with short, cone-shaped bill. Color of male: upperparts brownish-gray, grayer on crown and rump; rump and tail black, barred with white; face has white band bordered with black from base of bill to cheeks, which are orange; throat finely barred with pale gray and black, separated from white belly by black border; flanks chestnut, spotted with white; legs orange; bill and eyes red. Female lacks orange cheeks and throat bars.

THE ZEBRA FINCH IS AT HOME IN ARID AREAS. IT CAN *go for up to 250 days without water and can even drink quite salty water. The arrival of rain triggers courtship displays and nesting.*

DIET: seeds; insects in breeding season.

BREEDING: season variable, breeds after rain. Builds untidy domed nest of twigs, grasses and rootlets, lined with plant down, wool, fur and feathers, in low shrub or tree. Nest often re-used. Several pairs may nest in same tree. 3–7 pale blue or bluish-white eggs, incubated by female; male brings food. Chicks cared for by both parents; fledge at about 1 week. May have several broods each year.

OTHER INFORMATION: popular caged bird worldwide. Lives in large flocks.

VILLAGE WEAVER

Ploceus cucullatus

Black-headed weaver,
Spotted-backed weaver
Family: Ploceidae

DISTRIBUTION: forests, farmland and gardens of Africa, from Sudan and Ethiopia to Angola and the Cape.

SIZE: L 15–18 cm (6–7 in); WT 32.5–50.0 g (1.1–1.8 oz).

FORM: color of male bright golden yellow, tinged and mottled with dark brown on crown, nape, back and rump; wings and tail blackish, feathers edged with yellow; upper breast reddish in some races; black face mask, extending to throat; legs flesh colored; bill black; eyes red. Female similar, but upperparts dull brown, belly white; forehead, cheeks and nape yellowish-green, streaked darker; flanks grayish-brown; bill yellowish-brown.

THE MALE VILLAGE WEAVER BUILDS A HANGING NEST OF *woven stems, then hangs below it as he displays to the females, whirring his wings. Once he has mated, he abandons the female..*

DIET: seeds (including cereals); also plant juices.

BREEDING: breeds in colonies in same tree. Male builds hanging flask-shaped nest, hung from twig over water. Female lines it with grasses and other soft material. 3 greenish-blue eggs, speckled with brown or reddish-brown; incubated by female for 12–15 days. Chicks cared for by female alone. Fledge at about 10–15 days.

OTHER INFORMATION: forms large noisy flocks.

HOUSE SPARROW

Passer domesticus

Family: Passeridae

A HIGHLY ADAPTABLE OMNIVORE, THE HOUSE *sparrow has flourished almost everywhere that human habitation occurs.*

DISTRIBUTION: probably originally in open countryside and farmland from Europe and North Africa to western Asia. Introduced elsewhere.

SIZE: L 14–18 cm (5.5–7 in); WT 20.0–34.5 g (0.7–1.2 oz).

FORM: small bird with stout, conical bill. Male has rich brown upperparts, streaked with black; primaries and tail feathers blackish-brown, with pale edges; narrow white wingbar; crown gray; cheeks white; eyestripe, chin and throat bib black; underparts grayish-white; legs pale brown; bill black in summer, yellowish-brown in winter; eyes light brown. Female duller.

DIET: seeds, fruits, buds, flowers; insects, worms, human food waste.

BREEDING: pair build untidy globe-shaped nest of grasses, dry stems and debris, lined with feathers, with side entrance, in trees, bushes, inside other nests or in crevices in walls. 3–6 whitish or bluish-white eggs, marked with shades of gray and brown, incubated mainly by female for 9–18 days. Chicks fledge at 11–18 days.

OTHER INFORMATION: performs courtship displays in which several males display to and pursue a single female.

WHITE-HEADED BUFFALO WEAVER

Dinemellia dinemelli

Family: Ploceidae

DISTRIBUTION: dry thornbush, acacia woodland and scrublands of Africa, from southern Sudan and Ethiopia to Tanzania.

SIZE: L 23 cm (9 in); WT 57–71 g (2.0–2.5 oz).

FORM: medium-sized, sturdy bird with large head and large, thick bill. Color of upperparts black or dusky-brown; sometimes the primaries have white bases and the secondaries are white-edged; lower rump, upper and under-tail coverts and spots on elbow bright orange-red, which are conspicuous in flight; head and underparts whitish; legs brownish-black; bill blackish; eyes yellow or brown.

THE WHITE-HEADED BUFFALO-WEAVER BUILDS A COM-munal woven nest of thorny twigs, very hard for predators to penetrate. Each nest contains several nest chambers, one for each pair.

DIET: seeds, fruits; also insects.

BREEDING: season variable. Nests in scattered colonies in thorn trees. Male builds large, untidy hanging flask-shaped nest of twigs, lined with grasses and feathers, with entrance at bottom. 3–4 green or greenish-white eggs, marked with olive-brown, black and gray; incubated for 11 days. Chicks fledge at 20–23 days.

OTHER INFORMATION: calls range from harsh trumpeting sounds to bubbling twitters.

EUROPEAN STARLING

Sturnus vulgaris

Family: Sturnidae

DISTRIBUTION: open woodland, parks and gardens in Europe and western Asia; introduced to many parts of world.

SIZE: L 21 cm (8 in); WT 79.9–84.7 g (2.8–3.0 oz).

FORM: color blackish, iridescent, variously tinged with reddish-purple and purple; feathers of upperparts buff-tipped, those of underparts white-tipped, especially in winter; legs reddish-brown; bill greenish- or grayish-brown, yellow in summer; eyes dark brown; female more spotted, less iridescent.

DIET: insects and other invertebrates; also small vertebrates, plant material.

THE STARLING IS A COMMON BIRD IN URBAN AREAS. *Flocks of starlings feed on the ground, strutting around with a jaunty gait and probing the soil with open bills for insects and worms.*

BREEDING: in spring and summer. Builds untidy nest of grass and stems, lined with feathers, in cavity in wall, roof, cliff, rock, or in tree hole. Male starts nest before pairing; female finishes it. 5–7 pale blue eggs, incubated by both parents for 12–15 days, Chicks cared for by both parents. Fledge at 20–22 days, but are fed by parents for some time.

OTHER INFORMATION: chattering, whistling song often mimics those of other species.

HILL MYNAH

Gracula religiosa

Indian hill mynah,
Grackle
Family: Sturnidae

DISTRIBUTION: tropical forests up to
1,500 m (5,000 ft) of India, Sri Lanka
and southeast Asia, excluding
Philippines.

SIZE: L 29 cm (11.5 in); WT 161–229 g
(5.7–8.1 oz).

FORM: medium-sized bird with chunky
head and large, stout bill. Color black,
with glossy purple and green sheen;
broad white band on primaries; bare
skin and fleshy wattles on sides of head
and nape bright orange-yellow; legs
yellow; bill orange; eyes dark brown.

DIET: fruits (especially figs), berries,
buds, nectar; insects, small lizards.

THE HILL MYNAH IS A POPULAR CAGED BIRD BECAUSE
*it is a great mimic and can learn human
speech. Sometimes people put up nestboxes,
so they can collect chicks for the pet trade.*

BREEDING: April to July. Makes loose
nest of twigs, grass, debris and feathers
in tree hole, up to 15 m (50 ft) above
ground. 2 pale grayish or greenish
eggs, marked with brown, incubated by
both parents for 11–18 days. Chicks
cared for by both parents.

OTHER INFORMATION: forms very noisy
small flocks; wings make whirring
sound in flight. Usually lives in tree
tops; moves heavily on ground. Calls
made up of a mixture of whistles, war-
bles and shrieks.

MAGPIE-LARK

Grallina cyanoleuca

Mudlark, Peewee, Peewit
Family: Grallinidae

DISTRIBUTION: grasslands and farmland throughout Australia, including Tasmania; also found in New Guinea and Timor.

SIZE: L 27 cm (10.5 in); WT 89 g (3.1 oz).

FORM: color of upperparts mainly black, with broad white band from shoulder to inner wing; rump and underparts white; male has broad white eyebrow, large white patch on side of neck, black throat and upper breast; female has white forehead and throat, white neck patch extending behind eye and down to white flanks; legs black; bill yellowish-gray; eyes pale yellow.

MAGPIE-LARKS WORK TOGETHER TO REAR THEIR *young. They proclaim their territory by singing a "pee-wee" duet.*

DIET: mainly insects, cattle ticks, freshwater snails, other invertebrates.

BREEDING: July to January. Several pairs nest in same tree. Builds bowl-shaped nest of mud, mixed with wool and grass, lined with grass and feathers, on branch 9–12 m (30–40 ft) above ground, usually over water. 3–5 white or pinkish eggs, spotted and blotched; incubated by both parents for 17–18 days. Chicks fledge at 27–28 days.

OTHER INFORMATION: often have aerial fights with neighboring pairs.

BLACK-FACED WOODSWALLOW

Artamus cinereus

Family: Artamidae

DISTRIBUTION: dry savanna woodland, grasslands and other open country throughout Australia except extreme southeast and Tasmania; also Timor.

SIZE: L 18 cm (7 in).

FORM: small bird with heavy head and narrow, tapering wings and pointed bill. Color of upperparts dusky-gray, with bluish-gray wings; rump black; tail black with broad white tip; black band from eye around bill; underparts pale reddish-gray to buff or white, black under tail; legs bluish-gray; bill pale blue with black tip; eyes dark brown.

DIET: insects, caught in air or on the ground.

DESPITE ITS RATHER HEAVY APPEARANCE, THE BLACK-*faced woodswallow is an agile flyer, hawking for insects in the air like any true swallow. But it often prefers to wait on a perch, darting out to snatch passing insects.*

BREEDING: in spring in temperate regions, after rain in arid areas. Pair build cup nest of dry grasses, stems and rootlets, lined with grasses or hair. 3–4 white to bluish-gray eggs, spotted and blotched with reddish-brown and purplish-gray, incubated by both parents for 12–16 days. Chicks are cared for by both parents and fledge at about 16–20 days.

OTHER INFORMATION: often stand in close-packed rows on bare branches. Call a twittering "quet-quet".

GRAY BUTCHERBIRD

Cracticus torquatus

Family: Cracticidae

THE GRAY BUTCHERBIRD IS A *hunter of mice, lizards and small birds. Its feet are not strong enough to grip them while feeding, so it carries prey back to its perch and impales it on a thorn or wedges it between twigs while it feeds.*

DISTRIBUTION: in most habitats throughout Australia, except rainforest interiors and desert areas,

SIZE: L 26 cm (10 in); WT 74–110 g (2.6–3.9 oz).

FORM: medium-sized bird with short legs and strong, straight, finely-hooked bill. Color of upperparts dove gray to bluish-gray, with white rump; tail has white tip; head black; small white patch in front of eye; white band extends from sides of neck to breast; underparts white; legs grayish-black; bill bluish-gray with black tip; eyes reddish-brown.

DIET: small birds, chicks, lizards, mice, insects.

BREEDING: from July to January. Nest an untidy saucer, made of twigs and rootlets, lined with dry grasses and rootlets, in small tree up to 10 m (33 ft) above ground. 3–5 green to pale brown eggs, speckled and blotched with red, purplish- or chestnut-brown, incubated by female for 23 days; male brings food. Chicks fledge at 4 weeks.

OTHER INFORMATION: has flute-like warbling song, often sung as duet.

SUPERB BIRD OF PARADISE

Lophorina superba

Family: Paradisaeidae

A MALE SUPERB BIRD *of paradise in full courtship display. Once he has mated, he leaves the female to rear the family, and resumes his display to attract more mates.*

DISTRIBUTION: rainforest and forest edge in mountains of New Guinea, at 1,500–2,850 m (5,000–9,500 ft).

SIZE: L 25 cm (10 in); WT 60–93 g (2.1–3.3 oz).

FORM: color of male blackish, with iridescent turquoise-blue crown and breast shield; legs and bill blackish; eyes dark brown. Female has brown upperparts; underparts and head whitish, tinged rufous-brown on breast, finely mottled with grayish-brown to black; crown dark.

DIET: mainly fruits, berries; also seeds, insects, frogs, lizards.

BREEDING: males display in groups in tree tops. Female builds bulky nest of twigs in tree. 1–2 pale eggs, spotted, blotched or smudged darker, incubated for 17–21 days. Chicks fledge between 17–30 days.

OTHER INFORMATION: in display, breast shield fans out, feathers of crown and nape form overhanging umbrella-like hood and mouth opens to reveal greenish-yellow lining.

MAGNIFICENT RIFLEBIRD

Ptiloris magnificus

Albert riflebird, Prince Albert riflebird
Family: Paradisaeidae

DISTRIBUTION: rainforest of lowlands and foothills of northeastern Australia and on New Guinea.

SIZE: L 33 cm (13 in); WT 175 g (6.2 oz).

FORM: stout bird with heavy bill. Male has velvety, glossy plumage; crown, nape, throat, upper breast and central tail feathers black with iridescent bluish-green; throat and upper breast feathers scale-like, erected to form broad fan; underparts blackish, with bronze-green to deep red breast band; flanks have long plumes; legs and bill blackish. Female cinnamon-brown.

THE DISPLAY OF THE MAGNIFICENT RIFLEBIRD IS *designed to show off the brilliant throat and breast plumage. The bird raises its head and swings it from side to side while fanning its strangely-rounded wings.*

DIET: fruit; also berries, seeds, insects, frogs, reptiles.

BREEDING: male displays in groups in tree tops. Female alone builds tree nest and rears young. 1–2 pale eggs marked with dark scribbles, incubated for 17–21 days. Chicks fledge between 17–30 days.

OTHER INFORMATION: has clear call that carries long way through forest.

SPOTTED BOWERBIRD

Chlamydera maculata

Family: Ptilonorhynchidae

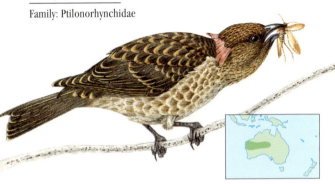

DISTRIBUTION: dense, dry woodlands, scrub and tree savanna of western and central Australia.

SIZE: L 28.5 cm (11 in); WT 128 g (4.5 oz).

FORM: a sturdy bird with a thick, slightly down-curved bill. Color brownish, with pale-tipped feathers giving mottled rufous and golden buff effect; has an iridescent rose-pink to magenta neck frill, smaller in female, that is erected like an inverted fan when displaying; tail tipped with orange-brown; legs olive-green; bill black; eyes brown.

DIET: fruit, berries, seeds, insects.

BREEDING: from October to December. Male builds bower up to 51 cm (20 in)

THE SPOTTED BOWERBIRD'S MOTTLED PLUMAGE *camouflages it in the dappled sunlight of the forest. To attract a mate, the male builds a double-walled bower of woven twigs, painted with reddish plant juices and decorated with white or green objects such as shells and bones.*

high and 76 cm (30 in) long, where he displays to females, erecting neck frill and making hissing and chirring sounds. Female makes shallow nest of twigs thinly lined with leaves and small twigs close to male's bower on branch, or in bush or mistletoe. 2 or 3 gray or greenish eggs with dark brown markings, incubated by female for probably 12–15 days. Chicks cared for by female; fledge at about 13–20 days.

OTHER INFORMATION: territorial during breeding season; forms small flocks at other times of year.

182

BLACK-BILLED MAGPIE

Pica pica

Magpie
Family: Corvidae

DISTRIBUTION: in woodlands and open country with scattered trees from tundra to semi-deserts throughout Eurasia, from western Europe and North Africa to Japan, and in temperate northwestern North America.

SIZE: L 45 cm (18 in); WT 135–209 g (4.8–7.4 oz).

FORM: a large crow-like bird with a long, tapered tail. Strikingly colored black-and-white, with black head, back, shoulders, tail and breast crescent; black areas have iridescent blue sheen; legs and bill black; eyes brown.

DIET: mainly insects; also other invertebrates, small mammals, baby birds, eggs, carrion, fruits, seeds.

THE BLACK-BILLED MAGPIE *is an inquisitive bird that is persecuted by gamekeepers for robbing bird nests. It likes to collect shiny objects, and is claimed to steal glass and jewelry.*

BREEDING: from early April. Pair build large untidy domed nest of twigs, lined with fine rootlets and hair, in bush or tree. 5–8 bluish-green or pale blue eggs, heavily spotted and speckled with olive-brown and gray, incubated by female alone for 17–18 days. Chicks hatch helpless, cared for by both parents. Fledge at 22–28 days.

OTHER INFORMATION: call a loud harsh chatter or chuckle. Walks or hops sideways on the ground.

SPOTTED NUTCRACKER

Nucifraga caryocatactes

Eurasian nutcracker, Alpine nutcracker, Nutcracker
Family: Corvidae

DISTRIBUTION: coniferous and mixed forests of northern and mountainous regions of Eurasia from Scandinavia, northeastern Europe, east to China and Japan. Occasionally spreads into temperate Europe.

SIZE: L 32–34 cm (12.5–13.3 in); WT 124–190 g (4.4–6.7 oz).

FORM: large bird with long, pointed bill and dark brown plumage covered in large white spots; sides of tail and undertail coverts white, which are conspicuous in flight; legs, bill and eyes blackish-brown.

DIET: nuts and seeds, especially pine seeds; some insects.

THE SPOTTED NUTCRACKER IS WELL CAMOUFLAGED *both among the conifer branches and on the ground. It feeds mainly on nuts and seeds, carrying them off in its throat pouch and burying them for use in winter.*

BREEDING: from mid-March. Nest a cup of twigs, moss and lichens with some mud and soil mixed in, lined with grass and lichens, high up against the trunk of a conifer. 2–5 pale blue or bluish-green eggs, finely spotted and speckled with olive-brown and gray, incubated by female alone for 17–19 days. Chicks cared for by both parents. Fledge at 21–28 days, but are not independent for a further 2–3 months.

OTHER INFORMATION: call a harsh "kraak". Flight undulating.

EURASIAN JAY

Garrulus glandarius

Jay
Family: Corvidae

THE EURASIAN JAY HAS HELPED TO SHAPE THE LAND-
*scape by burying acorns and other nuts in
open ground to eat later in winter. Nuts not
retrieved by the bird may germinate far
from the parent tree.*

DISTRIBUTION: woodlands, parks and
gardens across Eurasia, from Europe
and northwest Africa to Siberia, China,
Japan and southeast Asia.

SIZE: L 33 cm (13 in); WT 140–187 g
(4.9–6.6 oz).

FORM: large buffish to rufous or
brownish-pink bird with dark, sometimes grayish, shoulders; forehead
and crown paler with black streaks;
primaries and tail feathers blackish-
brown; wings have white and blue
patches; bill black; eyes bluish-white.

DIET: mainly nuts, seeds, fruits and
berries; also large insects, small mammals and young birds.

BREEDING: in April and May. Pair build
crude cup-shaped nest of twigs and
stems, in small tree. 5–7 pale bluish,
greenish or olive-buff eggs, finely spotted and speckled with olive or buff,
incubated by both parents for 16–17
days. Chicks fledge at 19–20 days.

OTHER INFORMATION: call a harsh
screeching "skaaak, skaaak".

ROCK DOVE

Columba livia

Feral pigeon, Pigeon
Family: Columbidae

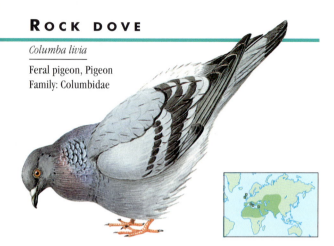

THE ROCK DOVE IS THE WILD FORM OF THE DOMESTIC *pigeon. It can fly at speeds of over 185 km (115 mi) per hour. Domesticated thousands of years ago, it was used as a messenger by the Romans.*

DISTRIBUTION: rocky cliffs and mountains across southern Eurasia from North Africa, western and southern Europe through Arabia and Asia Minor to the Indian subcontinent.

SIZE: L 31–34 cm (12–13 in); WT 340–369 g (12–13 oz).

FORM: stout bird with short neck, small head, short bill with fleshy "cere" at base around nostrils. Color variable, usually gray or bluish-gray with paler wings and back; wing-tips darker with two black wing-bars; rump usually white, visible in flight; iridescent green or purple patches at sides of neck; tail has black tip; legs purplish-red; bill black with whitish cere; eyes orange.

DIET: mainly seeds, grains, fruits and other plant material; also snails, insects, worms.

BREEDING: in spring and summer, in colonies. Nest a sparse platform of twigs, stems or roots. 1–3 white eggs, incubated by both parents for 17–19 days. Chicks fed at first on "pigeon milk", a curd-like substance secreted from lining of the crop and regurgitated. Independent at 30–35 days.

OTHER INFORMATION: call a soft repeated plaintive cooing.

VICTORIA CROWNED PIGEON

Goura victoria

Family: Columbidae

A MALE VICTORIA CROWNED PIGEON *performs a bowing display to a potential mate, fanning out its crest and tail feathers, booming invitingly.*

DISTRIBUTION: forest floors of the lowlands of northwestern New Guinea and nearby islands.

SIZE: L 66 cm (26 in); WT 2.4 kg (5.5 lb).

FORM: one of the world's largest pigeons. Color bluish-gray; lacy crest feathers fan-shaped and slatey-gray at tip, edged in white; crest permanently erected; tail feathers white-tipped; pale wing bar; lower neck and breast purplish-chestnut; black chin and mask around eyes; legs scaly, feet purplish-red; bill black; eyes red.

DIET: fallen fruits, seeds and berries.

BREEDING: nest a large platform of twigs and stems in a tree. 2 pale eggs, incubated for about 4 weeks. Chicks cared for by both parents. Fledge at about 3 weeks.

OTHER INFORMATION: forages in small flocks. Roosts in trees. Hunted for food.

CONSERVATION STATUS: very rare; threatened by hunting.

SCARLET MACAW

Ara macao

Red-and-yellow macaw
Family: Psittacidae

DISTRIBUTION: lowland tropical forests and open woodlands from southern Mexico to northern Bolivia and central Brazil.

SIZE: L 85 cm (33.5 in); WT 1.015 kg (2.2 lb).

FORM: large parrot with long, tapering tail. Color mainly scarlet; lower back, rump and tail coverts light blue; outer tail feathers blue; wing coverts yellow, tipped with green; undertail pale blue: legs purplish-gray; upper mandible pale horn color, lower mandible black; eyes pale yellow, surrounded by bare white face patch.

DIET: leaves, fruits and seeds; feeds high in trees.

THE SCARLET MACAW'S *brilliant plumage has been its undoing, attracting collectors and pet-owners. They are noisy in flight.*

BREEDING: nests in cavities in trunks of large trees. 2–4 china-white eggs, incubated by both parents for 24–25 days. Chicks fledge at 14 weeks.

OTHER INFORMATION: lives in pairs, family groups or larger flocks.

CONSERVATIONS STATUS: not at risk, but threatened by deforestation and by trapping for pet and collector trades.

HYACINTH MACAW

Anodorhynchus hyacinthus

Hyacinthine macaw
Family: Psittacidae

DISTRIBUTION: wet forest edges, palm plantations and watercourses in drier forests of eastern Bolivia, central Brazil and northeast Paraguay.

SIZE: L 95–99 cm (37.5–39 in); WT 1.5 kg (3.3 lb).

FORM: large parrot with long tail and long, narrow wings. Color deep cobalt-blue, with yellow chin and eye-patches; legs and bill grayish-black; eyes dark brown with yellow eye-ring.

DIET: nuts (especially palm nuts) seeds, fruits.

BREEDING: nest and unlined cavity in a tree trunk. 2–3 china-white eggs.

THE HYACINTH MACAW IS THE WORLD'S largest parrot. Like most parrots, it uses its feet to help manipulate its food. In terms of manual dexterity, parrots are unsurpassed by any other bird group.

Chicks incubated by both parents for about 27 days. Fledge at about 3 months.

OTHER INFORMATION: lives in pairs or family parties of 3–6 birds. Makes harsh screeches while in flight.

CONSERVATION STATUS: endangered by trapping for the pet trade and collectors, by hunting and through habitat loss due to clearance for agriculture and timber. Its survival may well depend on captive breeding.

CRIMSON ROSELLA

Platycercus elegans

Crimson parrot, Mountain lowry,
Red lowry, Red parrot
Family: Psittacidae

THE CRIMSON ROSELLA IS ONE OF AUSTRALIA'S MOST *colorful birds. It often visits parks and garden bird tables, but is also found high on mountain slopes. Young birds have striking green and red plumage; they travel around in very large flocks.*

DISTRIBUTION: rainforests and moist woodland and scrub, parks and gardens of eastern Australia up to 1,900 m (6,000 ft), excluding Tasmania.

SIZE: L 36 cm (14 in).

FORM: color rich crimson, with mottled black-and-red back; cheeks, shoulders, primaries, underwing coverts and tail violet-blue; legs gray; bill grayish-brown; eyes brown.

DIET: mainly seeds of grasses, eucalypts, acacias and other trees and shrubs; also fruits, nectar, insects.

BREEDING: from September to February. Nests in tree cavities. 5–8 white eggs laid on platform of decayed wood debris, incubated by female for about 3 weeks; male brings her food. Chicks cared for by both parents; fledge at about 5 weeks.

OTHER INFORMATION: lives in pairs or small groups. Has fast, undulating flap-glide flight. Stays in same area all year. Call a pleasant repeated three-note whistle; also harsh screeches and softer chatters.

KEA

Nestor notabilis

Family: Psittacidae

DISTRIBUTION: alpine grasslands, scrub and forest edges in the mountains of South Island, New Zealand, up to 2,000 m (6,500 ft). May move to coastal forests in winter.

SIZE: L 46 cm (18 in); WT 136–172 g (4.8–6.1 oz).

FORM: large, heavily built parrot with long, pointed upper mandible (longest in male), sturdy legs and large feet. Color iridescent green to olive-green, browner on breast and belly; feathers have dark edges; underwings bright tangerine, flashing in flight; legs pale grayish-brown; bill black; eyes brown.

DIET: mainly fruit, leaves, insects, grubs, carrion. Feeds on ground.

BREEDING: from July to January.

THE KEA IS A BOLD, INQUISITIVE GROUND-DWELLING *parrot. It is the parrot equivalent of a vulture, feasting on carcasses.*

Female makes nest of twigs, moss, lichens, leaves and chewed wood in rock cavity, under tree roots, on cliff ledge, in hollow log or on ground. Tunnel to nest chamber may be up to 7 m (23 ft) long. 3–4 white eggs, incubated by female for 17 or more days. Male brings food. Chicks cared for by both parents. Fledge at 14 weeks, independent 4–6 weeks later.

OTHER INFORMATION: call a harsh "kee-ah".

RAINBOW LORIKEET

Trichoglossus haematodus

Blue mountain lorikeet, Blue mountain parrot,
Rainbow parrot,
Blue-bonnet, Bluey
Family: Psittacidae

DISTRIBUTION: rainforests, eucalypt and other forests, plantations, heaths, scrublands, parks and gardens of coastal southeastern, eastern and northern Australia, New Guinea, and nearby islands of Pacific Ocean and Indonesia.

SIZE: L 28–32 cm (11–12.5 in); WT 122 g (4.3 oz).

FORM: color very variable; generally dark-green with purplish-blue crown and face, and yellow or red band at base of nape; breast and flanks orange-red; belly dark violet-blue; underwing coverts orange and yellow with blackish tips; underside of tail dirty yellow; legs greenish-gray; bill coral; eyes red.

THE MALE RAINBOW LORIKEET HAS MORE THAN *30 elaborate courtship and threat display dances involving bobbing, bowing and wing-fluttering to show off his plumage.*

DIET: mainly nectar and pollen, especially of eucalypts; also seeds, unripe fruits and insects.

BREEDING: season variable. Nests in tree holes, on platform of decayed wood debris. 2–3 white eggs, incubated by female for 26 days, male staying nearby. Chicks fledge at about 8 weeks.

OTHER INFORMATION: a noisy, showy bird that travels in flocks.

ECLECTUS PARROT

Eclectus roratus

Red-sided parrot, Rocky river parrot
Family: Psittacidae

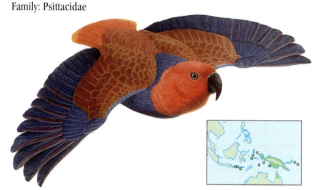

DISTRIBUTION: rainforests, moist eucalypt forests and scrub of northeastern
tip of Australia, New Guinea and nearby islands of the Pacific Ocean and
Indonesia.

SIZE: L 38–45 cm (15–18 in);
WT 428 g (15.1 oz).

FORM: plump parrot with short, square
tail. Color of male shiny green, with
patch of red on flanks and shoulders;
wingtips purplish-black; underwings
black; tail has yellow tip; legs dark
gray; bill large, upper mandible
orange, lower mandible black; eyes
orange. Female crimson red, with blue
belly, primaries and front secondaries;
tail has orange tip; underwings black;

*THE MALE AND FEMALE ECLECTUS PARROT ARE VERY
different. This female is bright red and blue,
while the male is brilliant green.*

underside of tail red; legs dark olive-
green; bill smaller, black; eyes yellow.

DIET: fruits, nuts, seeds, nectar, buds.

BREEDING: from September to
December. Colony nests in holes in tra-
ditional nesting tree up to 20 m (70 ft)
above ground, at forest edge. Nest hole
lined with decayed wood debris. 2 white
eggs, incubated by female for about 30
days; male brings food. Chicks fledge
at 8–9 weeks.

OTHER INFORMATION: form noisy, conspicuous flocks.

Sulfur-crested Cockatoo

Cacatua galerita

White cockatoo
Family: Psittacidae

Distribution: forests, grasslands and some urban areas of northern and eastern Australia, New Guinea and Melanesia, up to 1,500 m (5,000 ft).

Size: L 38–50 cm (15–19.5 in); WT 892 g (1.97 lb).

Form: large pure-white parrot with upswept sulfur-yellow crest and tinge of yellow on cheek, underwings and underside of tail; tail short, rounded; bill thick, powerful; legs short and feet strong. Legs and bill black; eyes dark brown to reddish-brown.

Diet: feeds on ground on seeds (including grains), fruits, buds, insects; uses bill to dig up roots and tubers.

LARGE, NOISY FLOCKS OF SULFUR-CRESTED *cockatoos whiten the ground as they descend to feed on seeds and fruits.*

Breeding: from May to January. Pair nests in unlined cavity high in eucalyptus tree, more rarely in holes in cliffs. Nest hole used all year. 2–3 white eggs, incubated by both parents for about 4 weeks. Chicks fledge at about 5–6 weeks.

Other information: erected crest is used for display, often while hanging upside-down.

BUDGERIGAR

Melopsittacus undulatus

Flightbird, Lovebird,
Zebra parrot,
Warbling grass parakeet
Family: Psittacidae

DISTRIBUTION: wooded country, scrub-lands, grasslands and cultivated land throughout most of Australia (excluding Tasmania), moving into coastal areas in spring and summer.

SIZE: L 17–20 cm (6.5–7.8 in); WT 28.8–29.2 g (1 oz).

FORM: small parrot with long, pointed tail. Color of wild birds green with bright yellow crown, mask and throat; black spots on throat; violet mustache; back barred with black; legs pinkish-gray; bill grayish-yellow; base of bill and fleshy cere blue in male, brown in breeding female; eyes yellow.

DIET: seeds of grasses, herbs and shrubs; also cereals.

BREEDING: season variable, according to rains. Many pairs nest close together,

THE WILD BUDGERIGAR IS THE *ancestor of the domestic bird. Courting pairs may croon to each other for up to 40 minutes in every hour hence the name "lovebirds".*

in holes in trees, fence posts and logs. Eggs laid on layer of wood dust. 4–6 white eggs, incubated by female for 18 days; male stays nearby. Chicks fed at first on a secretion from the female's digestive tract, then on seeds. Fledge at 4 weeks. Remain with parents for further 2 weeks.

OTHER INFORMATION: often forms huge flocks that wander over large areas in search of food and water. Lives in trees but feeds on ground.

EURASIAN CUCKOO

Cuculus canorus

European cuckoo,
Common cuckoo, Cuckoo
Family: Cuculidae

THE EURASIAN CUCKOO IS A NEST PARASITE: IT LAYS *its eggs in the nests of other birds and leaves them to rear its chicks. Some cuckoos even lay eggs that match the coloring and pattern of the host's eggs.*

DISTRIBUTION: breeds in open country with trees or shrubs from western Europe and North Africa to eastern Siberia, Nepal, China and Japan. Winters in bushy country in southern Africa and Asia.

SIZE: L 33 cm (13 in); WT 113 g (4 oz).

FORM: has short down-curved bill, pointed wings and long, broad, tapering tail. Male bluish-gray, with darker wingtips; tail has white spots and tip; underparts whitish with narrow black bars; legs yellow; upper mandible dark brownish-yellow, lower one greenish; eyes orange-yellow. Female browner on breast; occasionally chestnut all over.

DIET: caterpillars, worms and other small invertebrates.

BREEDING: from late April onward. Female lays 8–12 eggs, one at a time in the nest of a small insect-eating bird, often removing one of the host's eggs and eating it. Chick hatches after 12 days, then ejects host's eggs and chicks. Fed by foster parents; fledges at 17 days.

OTHER INFORMATION: named for "cuck-oo" call of male.

HOATZIN

Opisthocomus hoazin

Stinkbird
Family: Opisthocomidae

THE HOATZIN IS A POOR FLIER: IT USUALLY GLIDES *from tree to tree or clambers through the branches, using its wings for support. Young have long, movable first and second digits with claws that they use as hooks for climbing.*

DISTRIBUTION: wet forests of South America, from Guyana and Ecuador to Brazil, Bolivia, and the Orinoco and Amazon river basins.

SIZE: L 60 cm (23.5 in); WT 855 g (1.9 lb).

FORM: large bird with heavy body, large tail, long neck and small head surmounted by large spiky crest. Color of upperparts dark brown, streaked with white on neck and shoulders; outer primaries reddish; tail tipped with yellow; crest reddish-brown; legs dark; bill horn-colored; eyes red; bare skin on face blue.

DIET: leaves of arum and other swamp plants. Has huge muscular crop with horny ridges for grinding plants.

BREEDING: up to 6 adults may help build nest and care for eggs and young. Nest a frail platform of sticks in a tree above water. 2–5 yellowish-buff to creamy-white eggs, spotted with blue/violet or brown, incubated for about 28 days. Young soon leave nest, but are fed by both parents. If danger threatens, they will drop into the water.

OTHER INFORMATION: has strong unpleasant odor, so is not hunted.

OILBIRD

Steatornis caripensis

Guacharo, Diablotin
Family: Steatornithidae

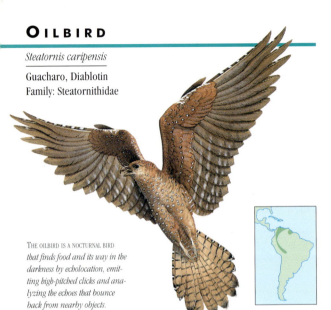

THE OILBIRD IS A NOCTURNAL BIRD
*that finds food and its way in the
darkness by echolocation, emit-
ting high-pitched clicks and ana-
lyzing the echoes that bounce
back from nearby objects.*

DISTRIBUTION: roosts on ledges in caves in humid forests, from northwestern South America east to the Guianas and Trinidad.

SIZE: L 43–48 cm (17–19 in); WT 375–480 g (13.2–16.9 oz).

FORM: has strong, hooked bill surrounded by sensitive bristles up to 5 cm (2 in) long, for feeling way in dark; large, reflective eyes for night vision; wings long, narrow; tail long; legs short and weak. Color rich brown, barred with black or dark brown, with scattered white spots, especially on outer edges of wings and tail; underparts cinnamon-buff; legs whitish-pink; bill dark yellowish-brown; eyes deep red.

DIET: takes fruits from trees while flying; stores them in its stomach and digests them in cave by day.

BREEDING: nonseasonal. Nest a mound of regurgitated seeds and droppings on a cave ledge. 2–4 white eggs, incubated by both parents for 32–35 days.

OTHER INFORMATION: forms large colonies of hundreds of birds.

198

GIANT HUMMINGBIRD

Patagona gigas

Family: Trochilidae

THE GIANT HUMMINGBIRD IS THE LARGEST *hummingbird in the world. It hovers more slowly than other hummingbirds (only 8–10 beats per second) while it feeds on nectar from deep, trumpet-shaped flowers.*

DISTRIBUTION: riverside forests, scrub and cultivated land in the Andes of South America, from Ecuador to Chile, and in lowlands of western Argentina. May move lower down the mountains during the winter.

SIZE: L 19–20 cm (7.5–8 in); WT about 20.2 g (0.71 oz).

FORM: slender bird with long, pointed wings, long tail and very long bill. Color greenish-brown on upperparts, rich reddish-brown on underparts, this color extending to eyes and nape of neck; base of tail whitish.

DIET: nectar, pollen, small insects.

BREEDING: female makes small cup-shaped nest of moss and lichens, on a branch or cactus stem. 2 white eggs, incubated by female for 14–19 days. Young cared for by female, who feeds them on regurgitated nectar and small insects. Fledge at around 19–25 days.

OTHER INFORMATION: being larger than other hummingbirds, this species has a lower surface area to volume ratio, which helps it to conserve heat in the cooler temperatures of the mountains. In direct flight it can reach speeds of 65 km (40 mi) per hour.

SWORD-BILLED HUMMINGBIRD

Ensifera ensifera

Family: Trochilidae

DISTRIBUTION: humid mountain forests and shrubby areas of the Andes, from Venezuela to northern Bolivia, up to 3,000 m (9,850 ft).

SIZE: L 25 cm (10 in); WT 12 g (0.42 oz).

FORM: has long, slender bill; long, slightly-forked tail; short legs; tiny feet. Color of male shining green, with coppery-green head, tail and wings; throat blackish; breast and sides of neck brilliant green; rest of underparts grayish-buff; legs and bill blackish-brown; eyes dark. Female less coppery on head; underparts green, feathers with white edges; throat bronze, feathers with wider white edges.

THE SWORD-BILLED HUMMINGBIRD HAS THE LONGEST bill of any hummingbird. It feeds at hanging, trumpet-shaped flowers. As it feeds, pollen rubs off on its head, to be deposited on the next flower visited, so pollinating it.

DIET: nectar and pollen; also catches insects in midair.

BREEDING: from early spring. Female builds deep mossy cup nest among roots of epiphytic plants on trees high above ground, anchored by spider webs and lined with plant material. 2 white eggs, incubated by female for 14–23 days. Chicks cared for by female alone. Fledge at 18–30 days.

OTHER INFORMATION: may become torpid at night to conserve heat.

RESPLENDENT QUETZAL

Pharomachrus mocinno

Quetzal
Family: Trogonidae

THE MALE RESPLENDENT QUETZAL *has two long, streamer-like tail feathers up to 1 m (3.3 ft) long, with a train of smaller ones, which ripple and flutter in flight, and especially in its aerial courtship displays.*

DISTRIBUTION: cloud forests up to 3,000 m (2,750 ft), from southern Mexico to Panama.

SIZE: L 35–38 cm (14–15 in); WT 206 g (7.3 oz).

FORM: male has erectable crest. Color of male brilliant iridescent green; lower breast, belly and undertail feathers crimson; wing primaries black, covered by long, golden-green wing coverts; upper tail feathers golden-green, tinged with blue or violet; outer tail feathers white; legs blackish; bill yellow; eyes black. Female similar but with smoky-gray head, breast and lower part of belly, and much shorter tail, barred black-and-white; bill black.

DIET: mainly fruit taken on the wing; also small insects and other animals.

BREEDING: from April onward. Pair excavate hole in rotting tree trunk up to 18 m (60 ft) above ground. 2 pale blue eggs, incubated by both parents for 17–18 days.

OTHER INFORMATION: quetzal feathers were a sign of rank in Aztec and Maya culture.

CONSERVATION STATUS: threatened by deforestation and by hunting for the cage-bird trade.

TOCO TOUCAN

Ramphastos toco

Family: Ramphastidae

DISTRIBUTION: open forests, woodlands and plantations of eastern South America, from Guyana to northern Argentina.

SIZE: L 61 cm (24 in); WT 540 g (1.2 lb).

FORM: large bird with very deep bill up to 19 cm (7.5 in) long and short, rounded wings. Its toes are arranged in two pairs. Color black, with white throat; legs dark gray; bill orange with black band at base and black patch at tip; eyes dark, surrounded by patch of yellow bare skin.

DIET: mainly fruits and small vertebrates; also insects and spiders.

THE TOCO TOUCAN IS THE WORLD'S LARGEST TOUCAN. *With the tip of its huge bill, the toucan can pick fruits at the tips of twigs too thin to take its weight. It then flings its head back and opens its bill, so the fruit falls into its mouth.*

BREEDING: nests in tree holes, often "borrowed", lined with wood chips. 2–4 white eggs, incubated by both parents for at least 16 days. Chicks cared for by both parents. Fledge at about 50 days.

OTHER INFORMATION: long beak is hollow structure of light-weight horn supported by internal struts; used to threaten rivals or enemies, as dueling weapon in play and in pushing contests. Lives in flocks of up to 12 birds.

RHINOCEROS HORNBILL

Buceros rhinoceros

Family: Bucerotidae

THE RHINOCEROS HORNBILL HAS A HUGE HORNY *casque on the top of its bill, which resembles the horn of a rhinoceros, but is actually used to knock down fruit.*

DISTRIBUTION: forests of lowlands and foothills in Malaya, Sumatra, Java, Borneo.

SIZE: L 120 cm (47 in); WT 2.5 kg (5.5 lb); male larger than female.

FORM: large bird with broad wings, long tail and very large down-curving bill with casque. Front 3 toes webbed at base. Color black, with white belly; tail white with black band; legs dark gray; bill yellowish to ivory; casque yellowish in front, reddish behind; eyes yellow, with orange eye-ring.

DIET: mainly fruits (feeds like toucan, using tip of bill like pincers); also small birds and other small vertebrates.

BREEDING: nests in tree holes. Female seals herself inside, using mud mixed with saliva, leaving small hole for bill to protrude, for defense against predators. 1–6 white eggs, incubated by female for 28–40 days; male feeds her and chicks, which fledge at 4–8 weeks.

OTHER INFORMATION: lives alone or in pairs. May forage in small groups.

RUFOUS-TAILED JACAMAR

Galbula ruficauda

Family: Galbulidae

THE RUFOUS-TAILED JACAMAR BEHAVES LIKE A LARGE
*flycatcher: it waits on a perch for insect prey
to pass by, then flies out to catch it, returning
to the perch to kill and eat it.*

DISTRIBUTION: one population from
Mexico south to northern Brazil and
northwest Ecuador, and a second from
central Brazil south to northeast Argentina. Found in moist tropical forest
clearings and in woodland beside water.

SIZE: L 30 cm (12 in); WT 58–76 g
(2–2.7 oz).

FORM: slender bird with long, tapering
tail, short legs and long thin bill. Color
of upper parts metallic green; outer tail
feathers rufous-cinnamon; throat white
in male, buff, cinnamon or black in
female; belly chestnut or buff; legs
grayish-yellow; bill black; eyes dark.

DIET: flying insects, especially butter-
flies, beetles and dragonflies. Beats
them against branch to kill them.

BREEDING: pair excavate tunnel 30–
45 cm (12–18 in) deep in bank or
earth or sand. 2–4 white eggs, incubated by both parents for 19–21 days.
Young cared for by both parents. Fledge
at 20–26 days.

OTHER INFORMATION: lives in pairs.

RED-HEADED BARBET

Eubucco bourcierii

Family: Capitonidae

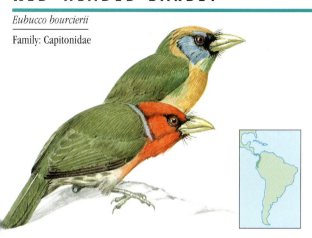

DISTRIBUTION: tropical mountain forests of Central and South America, from Costa Rica south to northern Peru and western Venezuela, usually at about 1,200–2,400 m (4,000–8,000 ft).

SIZE: L 15 cm (6 in); WT 33.5 g (1.2 oz).

FORM: small, plump bird with short neck, large head and large, heavy bill fringed with bristles. Male dull green, with glossy scarlet head and throat, fading to orange on breast; white half-collar on nape of neck; underparts whitish streaked with green; legs dark gray; bill yellowish-green; eyes brown. Female green with blue cheeks, and thin orange-yellow breast-band.

A PAIR OF RED-HEADED BARBETS SHOW THE FRINGE OF *bristles at the base of the bill that gives them their name. They have a variety of strange calls, ranging from rattles and chatters to toad-like trills.*

DIET: mainly insects; also fruits and flowers. Probe leaf litter and dead leaf clusters for insects; may hang upside-down to feed on fruits.

BREEDING: nest a layer of wood chips in a tree hole. 2–5 white eggs, incubated by both parents. Chicks fledge at 31 days, but remain with parents for a long time afterward.

OTHER INFORMATION: unlike most barbets, is solitary and usually silent. Sometimes joins mixed flocks.

INDEX

Artists

Norman Arlott
Trevor Boyer
Ad Cameron
Robert Gillmor
Peter Harrison
Chloë Talbot Kelly
Sean Milne
Denys Ovenden
Laurel Tucker